Thin Places

and Five Clues
in Their Architecture

C. Page Highfill, AIA Emeritus

*Thin Places - where the veil
between this world and
another becomes thin.*

WHAT OTHERS ARE SAYING...

As I read Page Highfill's book, I kept having flashbacks to the Thin Places in my own life -- the intimacy of friendship, shared laughter and shared tears, the relief of a diagnosis, times of spiritual awareness. I have stood at my parents' graves on a green hillside in Tennessee. I have stood before a stone cross in the coliseum in Rome. I have stood at the end of the rail line on the selection platform in Auschwitz -- Thin Places all. Read this book. It will make you think. It will make you remember. It will set your spirit free to imagine.
  -- Roberta M. Damon PhD.

Fascinating. Easy read prompting deep, enriching thoughts. Intriguing, relaxing and inspirational. One of Page's many captivating expressions in the book was from a mountain overlook in Virginia where he wrote, "...where God's beauty is the choir and His silence is our sermon."
  A most interesting subject. Truly, I will employ some or all of the five clues of Thin Places when I travel to some of our favorite places – and new places.
  -- Rev. Tom Lacy

As has been eloquently described by Page Highfill, in the Celtic tradition "thin places" are the physical settings that provide an accessible connection between the mortal world and the cosmic world beyond. As he has noted, both natural and man-made settings can be experienced as *thin places* and he has advanced the effective discussion of these places / spaces by proposing a tabular means of evaluating the characteristics of thin places that can provide a common ground for understanding, discussion, and perhaps, intention. In fact, he has extended the conversation about thin places in the very hope that it will help enable the intentional creation of architecture that can function in this way.
  My forty years or so of architectural practice could be characterized as an ongoing "search for sacred space" wherein we have sought to create spaces / places that are transformational settings that will enable people to reach the highest dimension of

their spiritual existence, to experience in mystical awe the ineffable and the sublime. In essence, we have sought to create architectural *Thin Places* that can consistently engender *Thin Moments*. I believe that architectural *Thin Places* can only be created when there is an intention to create them, and, most importantly when the Holy Spirit is specifically included in the creative process. Page's book is a great addition to the professional toolbox used in this effort because it can be used to facilitate communication about the meta-physical dimensions of design (the symbolic and the experiential) which is often left to the wayside because the discussions of the physical dimensions (the formal and the technical) are so much easier to discuss.

> *-- Wes McClure, FAIA Architect*
> *Raleigh, North Carolina*

I have learned a tremendous amount about architecture in Thin Places and now look at places with a more discerning eye. Indeed, Page has much to say in this book that will help people better understand their world and how they can fit into it with harmony and peace. This is an important book for those who are seeking to make sense of their environment.

> *-- Carol McLaren*

The description of a Thin Place, "where the veil that separates heaven and earth is lifted and one is able to receive a glimpse of the glory of God," is what Page so beautifully expressed in his book. I recently went to Asheville, NC and as I stood on the hillside near THE COVE, I suddenly realized I was seeing God's creation at its finest right here in the United States. After reading this book, I realize that God's glory is revealed to us many times in our life. We just need to relax the clamps we have on earth and let the veil give us that glimpse of heaven and the glories that are awaiting us. My travels now will be richer because I have a better understanding of Thin Places and more importantly, how to enjoy them to the fullest.

> *-- Joyce Clemmons*

Thin Places
and Five Clues within Their Architecture

C. Page Highfill, AIA Emeritus

Thin Places – where the veil between this world and another becomes thin

Thin Places and Five Clues in Their Architecture
by C. Page Highfill, AIA Emeritus

EnterPaths, L.L.C., PUBLISHER
12915 River Road
Richmond, VA 23238

Orders and/or information: http://www.EnterPaths.com/

ISBN 978-0-578-03158-3

This book is dedicated to the
Holy Cosmic Truth, known by many
names, nods and nurtures. While I choose
one of them, I dedicate this book to all of them: the
whole; the universally intrinsic; the ways and Thin Places,
which seek to ennoble for all – glimpses of the Divine.

This book is also dedicated to my family: my wife, Kate;
our four children, Scott, Marc, Ann Page and Katie;
and their spouses Cindi, Sharon, Rob and Bruce; and
their children, Bryan, Holly, Sean, Austin, Matt, Riley,
Davis, Kyle, Jake, Sydney, Brooke, Rebekah, Olivia and Clare.

This book is also dedicated to Hampden-Sydney College,
Hampden-Sydney, Virginia, and to its leaders, administrative
and facility staff, professors and students, with whom
we all worked and shared treasured opportunities for
nearly twenty years to further her noble mission.

CONTENTS

TO SCORE PLACES YOU VISIT

Blank Score Sheets with Response Subjects
Blank Score Sheets without Response Subjects
Quick Order Forms for Information and Ordering

Disclaimer – Warning

This book is designed to provide information on Thin Places and certain common characteristics the author has perceived over the years as five design clues within their architecture. This book is sold with the understanding that the publisher and author are not engaged in rendering architectural, legal, accounting or other professional services here. If legal or other expert assistance is required, the services of a competent professional should be sought. Readers are encouraged to involve their own architects and building team in any specific applications of the information contained herein.

It is not the purpose of this book to reprint all of the information that is otherwise available on this subject, but instead to complement, amplify and supplement other texts. You are urged to read all the available material, learn as much as possible about Thin Places and tailor the information to your individual and project needs. For more information, see the various web sites available from search engines responding to a Thin Places search.

Designing Thin Places or other highly engaging places is not a simple or quick process. Qualified competent certified architects should lead the design efforts in concert with clients, building owners and facility sponsors, all working together as a team. Sharing the information in this book within your professional team is encouraged as the most effective way to move forward and to fruition some of the benefits noted herein.

Every effort has been made to make this book as complete and as accurate as possible. However, there may be mistakes, both typographical and in content. Therefore, this text should be used only as a general guide, and with qualified professionals as noted above.

The purpose of this book is to educate, inspire and entertain. The author and the publisher shall have neither liability nor responsibility to any person or entity with respect to any loss or damage caused, or alleged to have been caused directly or indirectly, by the information contained in this book.

If you do not agree with the above, you may return this book to the publisher for a full refund.

1

INTRODUCTION
What this is all about

"Thin Places" date back many centuries. One turning point in their history occurred around 500 B.C. when the Celts migrated to and began building monuments and sacred sites in Scotland, Ireland, England, Northern Ireland and Wales. Almost a thousand years later, St. Patrick introduced Christianity to the Celts around 431 A.D., and they converted from their pagan religions to Christianity. Thin Places continued to be built and included as important parts of early Celtic culture and in Christian rituals. A Thin Place is customarily defined by anthropologists, historians, photographers, sacred site enthusiasts and others as a place where the veil between this world and another world becomes so thin that it nearly disappears. They are places where people have reported experiences and connections between spirit and matter, places where times of different eras often seem to merge, and places where people feel strongly connected with a divine presence. Stonehenge is considered a Thin Place by many. The Taj Mahal in Agra, India, has been described by many visitors as a Thin Place. Walking Labyrinths (as from Celtic history) may have been originally designed to be Thin Places. We will review these in more detail later.

A Google search of Thin Places performed during this writing yielded over 600,000 hits in .28 seconds. One of my favorite Thin Places websites is by Mindie Burgoyne and located at ://www.writingthevision.com/thinplaces.htm. I suggest you explore some of the many web sites and read the reactions of observers. This book is a supplement to other information, books, articles and websites on Thin Places. My specific focus is on *what can we learn from those special places* – what patterns might we find in their design which empower much higher human engagements and spiritual connections than at non-thin or thick places? Can we build upon those patterns and assemble design criteria of high engagement or of Thin Places? Can Thin Places actually be planned and created at new sites? What might be the economic, spiritual, biological and other benefits by including new Thin Places in our environments? How?

For whom is this book written?

If you have a keen sense of curiosity about these ancient and contemporary places; or if you are a spiritual dreamer and one who is sensitive to God's creation and humanity's co-creation, wondering how these Thin Places might work or wondering why so many people worldwide have reported very similar experiences at such places – this book is for you. I will explore with you in a step-by-step process the Five Clues I consider as probable answers to some of those thoughts. Over the years, I have found these Five Clues to be present, some very intensely, within the architecture of Thin Places, and I believe they are an important basis through which those places continue to generate such positive human engagement and repeat visitation. People just want to be there. They are empowered, inspired, many moved to tears, and deeply touched just by being there, and many keep coming back.

From a practical standpoint, this book is also written for those of you who make important environmental choices personally and for others in facilities for living, working, learning, worshiping and playing. Perhaps you are an owner, officer or board member, or a trustee or building committee member for new or renovated building projects. This book will explore some new place-engaging and cost-effective criteria to program into your facilities.

If you are an architect, this book reinforces your valuable efforts and some of what you may have known intuitively through art forms for years. This book will also explore with you the Five Clues in the architecture of Thin Places, two of which are rarely if ever taught in architectural schools and thus may be a totally new consideration for you and your clients. This book will also enable your clients and committee members to share in and understand some of the depth of design skills, experiences and realizations on paper and on screen of what you are discovering and moving towards in their behalf.

If you are a builder, or developer, individual or corporate owner of facilities who use the services of an architect, this book may help expand your understanding of what architects seek in everyone's behalf. Thin Places and highly engaging places do not necessarily cost more. Often they cost less to build and often substantially less to operate.

Perhaps you are one who in the past has chosen to utilize "layout-build" procedures as a streamlined non-architect application of the design-build process. If so, this book is also written for you. Learning about the engaging power of Thin Places may help you and those involved with you to recognize and understand the potential human influence of highly engaging facilities. This includes considering how to leverage long term economic feasibility by linking with the hearts and spirits of its tenants, customers, pupils, residents and others through the magic of art forms and these Five Clues. Such a connection rarely happens by accident and virtually never occurs through processes which intentionally exclude those benefits.

Let's look at two Thin Place examples.

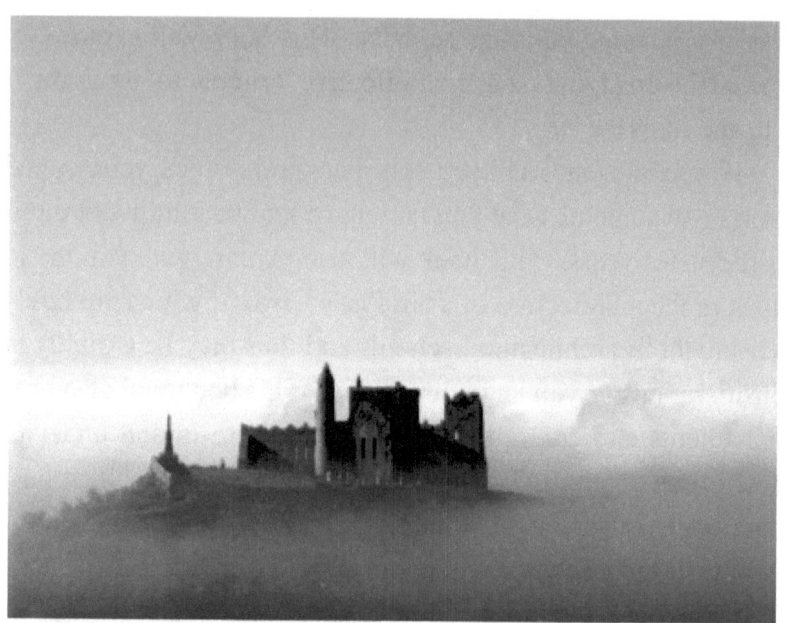

The Rock of Cashel, Tipperary, Ireland (Getty Images)

Ireland houses hundreds of Thin Places. One of the most frequently visited is the Rock of Cashel, which dates back to the 4th Century A.D. Many visitors report that Cashel is one of the "thinnest" of all Thin Places in the World. What engaging secrets might it hold for new projects?

Mindie Burgoyne, in her previously mentioned "Writing the Vision" website said of Cashel: "With every visit, Cashel is different, yet the same. Every departure from Cashel is painful and every return natural, as if there is a small presence that welcomes only me." Mindie also said at the same website: "Truth abides in Thin Places; naked, raw, hard to face truth. Yet the comfort, safety and strength to face that truth also abide there. Thin Places captivate our imagination, yet diminish our existence. We become very small, yet we gain connection and become part of something larger than we can perceive."

Another very special place is much newer and located in the United States is Thorncrown Chapel, a low-budget wood and glass structure designed by Fay Jones, FAIA, in Eureka Springs, Arkansas, winner of two national AIA awards; it is also fourth on the AIA's list of top buildings of the twentieth century. <u>Thorncrown Chapel continues to draw so many visitors and patrons that its modest tourist fees add up annually to more than its original cost.</u> I consider Thorncrown Chapel a 20th century Thin Place. We'll look at the details of "why" later. Understanding some of its brief history now will help you see how its background also contains a common thread with many other Thin Places.

The following is quoted by permission, from the website, (11-14-06) .thorncrown.com/history.htm. "Thorncrown was the dream of Jim Reed, a native of Pine Bluff, Arkansas. In 1971 Jim purchased the land which is now the site of the chapel to build his retirement home. However, other people admired his location and would often stop at his property to gain a better view of the beautiful Ozark hills. Instead of fencing them out, Jim decided to invite them in. One day while walking up the hill to his house, the idea came to him that he and his wife should build a glass chapel in the woods to give wayfarers a place to relax in an inspiring way.

Shortly thereafter Jim met E. Fay Jones, Architect FAIA and a professor at the University of Arkansas at Fayetteville. Much to Jim's surprise, Jones was quick to accept the proposal to design the chapel. On March 23, 1979, the construction crew broke ground on the mountainside. Jim's dream looked as if it would soon be a reality.

However, halfway through the project funds began to run out. Soon the building process ground to a halt. In his own words, it looked like Jim had made "...the biggest mistake of my life." He desperately tried to raise the necessary funds to complete his dream, but all of his efforts failed.

Finally, one evening Jim took what he thought would be one last walk down to his half-finished chapel. He would take one last look and never return. Then the unexpected happened.

He said, "I am not proud of the fact, but the first time I ever got down on my knees was on the chapel floor. I prayed more seriously than ever before. All the trials and tribulations gave me the humility to get on my knees."

This was a turning point in Jim's life and in the construction of the chapel. In a few short days all the money Jim needed was made available. On July 10, 1980 Thorncrown Chapel opened. Since then over five million people have visited this little chapel on the hillside. Thorncrown has won numerous architectural awards. It has been featured on television programs such as NBC Nightly News and the 700 Club. Almost every major magazine in the country has carried a story about the chapel including *Time, Newsweek, and Parade.*"

A very well-done panorama virtual tour of the interior of Thorncrown may be viewed at this website:

://host.newspin360.net/thorncrown/pano.html

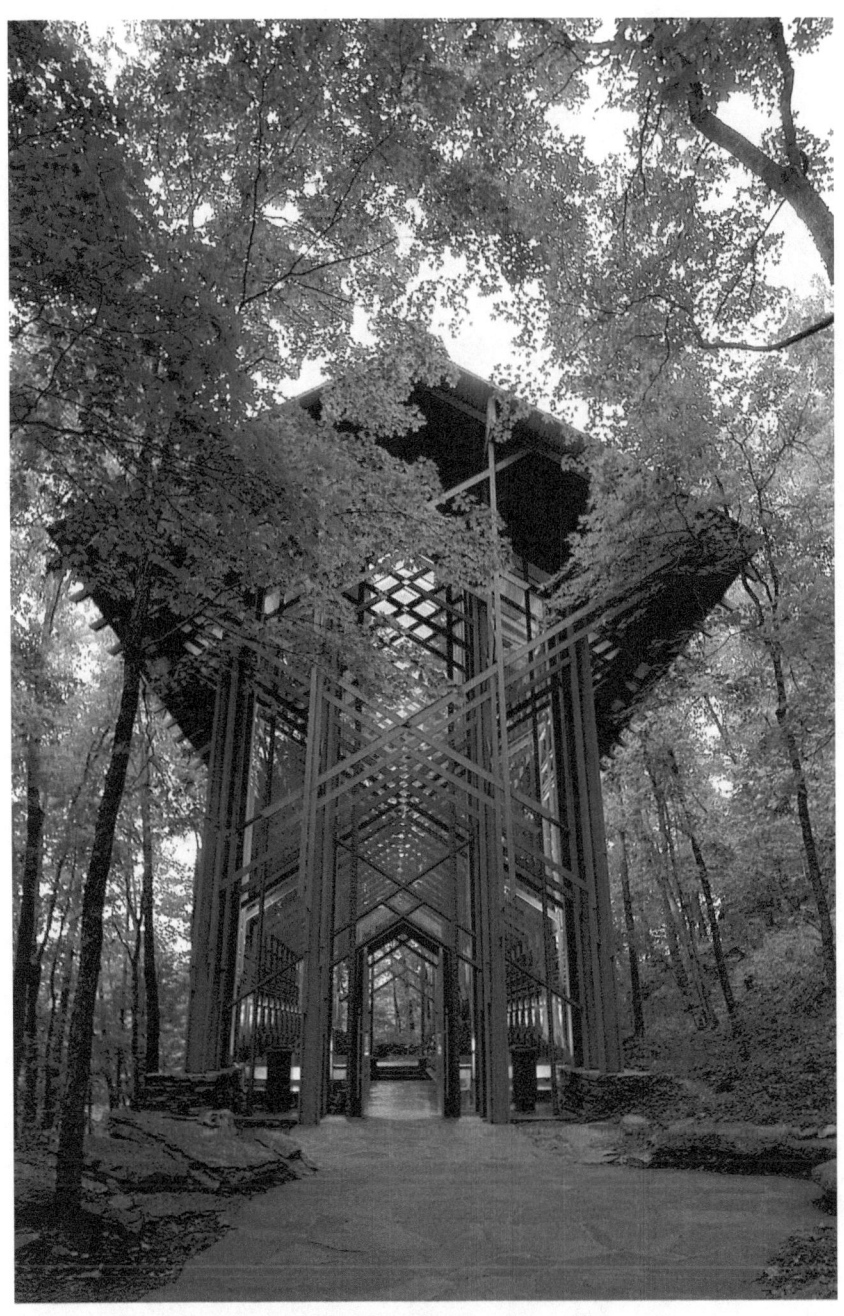

Thorncrown Chapel, Eureka, Arkansas
Photo by Whit Slemmons

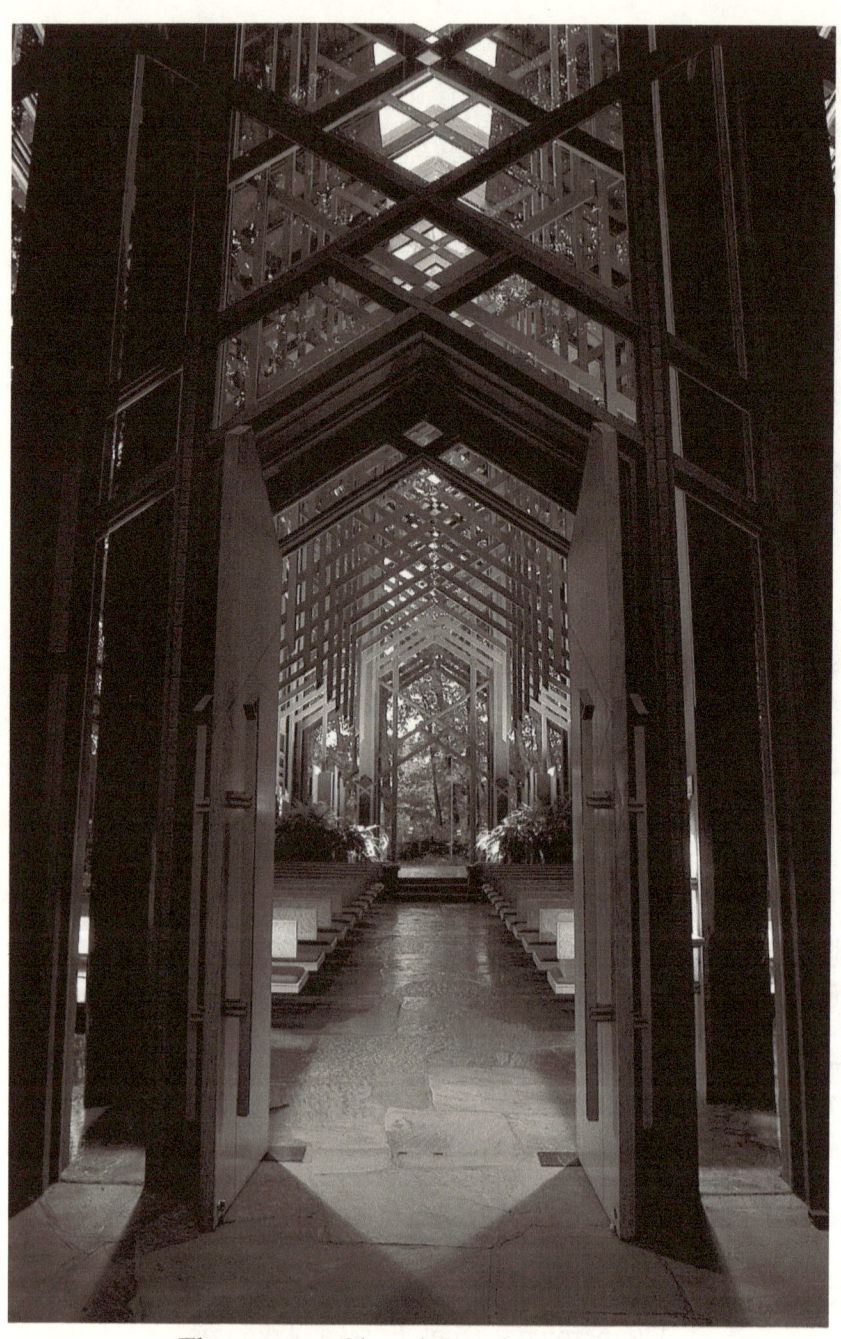

Thorncrown Chapel, Eureka, Arkansas

Photo by Whit Slemmons

We'll look at Thorncrown Chapel in more detail later. Now consider the powerful drawing force behind this low budget project. How might we capture some of that engaging potential?

What can we learn from Thin Places?

As an architect, I have experienced many Thin Places in Virginia, across the USA mainland and on the Hawaiian Islands. I later write about two of them I experienced as a teenager and which ultimately set me on the path to a career in architecture.

Usually a Thin Place reveals itself initially to me through some form of intense beauty: a stunning sunrise at the seashore, an old historic building in the moonlight or the view from mountain overlooks above fog-covered valleys.

The intense beauty captures me. If this is a Thin Place, there's much more here than simply a "pretty scene." If I linger a bit, I might start to feel that there is something very special here. I may sense that I have been here before, even if I haven't. I may lose my sense of time, and I may even sense that I knew some of the people who were a part of this place previously.

If some of your experiences at special places are anything like mine, you are not sure what it is but you may begin to feel that you just ran into an old friend. You also might feel like you have been there before, even though you have not. Sometimes you may start thinking about other families who might have lived or died here. The place continues to welcome you, and you feel like you would prefer to just hang around for a while longer and absorb the spirituality that surrounds you. You may nod to yourself and say, "Yes, there's something very unusual about this place." If your spiritual senses are awakened, you may wish to simply pause and meditate in prayer for a few moments. You may even feel as though you are walking within a prayer.

These are similar feelings I had as I stepped inside one of the James River plantation chapels along Route 5, located between Richmond, and Williamsburg, Virginia. I was not looking for, nor did I expect this old weathered chapel to be a Thin Place. But it was to me on that day. It was (clue one) a History Dense place. Suddenly I felt as though I had been there before (but had not) and I sensed the presence of the people and families who once worshiped there.

Why is poetry included in this Book?

Imagine for a moment that you meet a stranger at a social event. The two of you talk, and soon both of you discover that you have a mutual friend. The conversation continues, now reflecting memorable experiences each of you recall about that friend. You continue to talk, and discover more mutual friends and experiences. Soon, you are no longer total strangers. You have gotten to know a little bit more about each other through both of you knowing mutual friends.

You and I are essentially strangers. And, while we may not know any mutual friends, we do have one: the language of poetry. Poetry can more accurately communicate the feelings we experience – such as at Thin Places – than any other form of language. During the writing of this book, I shared portions of my work with friends and colleagues. I was surprised when many of them later said to me, "I remember having feelings like you described in some of your poems, but I just never conveyed those to others before." Accordingly, I hope some of the poems I have included about my Thin Place experiences over the years will be received by you as friends we both share.

Later in my career, I noticed that architectural communication and impressions at Thin Places appeared to come through and from different sources than most normal architectural settings, and tended to be exaggerated and intensified, speaking at much higher concentrations. I pursued and wrote about those differences for years. I wanted to deconstruct those dissimilarities in order to see if there might be some common denominators, like "fingerprints" of Thin Places. If common denominators could be identified, they may present new opportunities to learn and expand our skills to provide and manage more meaningful and engaging environments in the future.

What are the five clues?

The Five Clues are as follows: History Density, Cultured Symbols, Intrinsic Symbols, Bio-Kinetics, and Intention. We will review each of these in detail later. We will also review ways I recorded my perceptions at various Thin Places on a score sheet. Scoring Thin Places with respect to the content and intensity of each of the Five Clues, revealed some interesting patterns I had not discerned previously. I have included some blank score sheets at the end of the book that you can use to score your own perceptions. I think you will find the process quite enlightening.

These five clues, of course, are not the only elements in the architectural chorus, but I have found these to be the key ones at Thin Places. The five are also found in varying degrees in other architectural examples as well, either at a positive level or more often at a deficient or negative level. Relationships between various score sheets of numerous places suggested a tendency to me: the higher the level and intensity of certain perceived clues, potentially the thinner and more engaging the place may be. The reverse was true as well. That is: the more negative (absence or lacking) of certain clues, the thicker and less engaging the place may be.

Obviously the design skills of the place designers could alter this tendency either way. For example, an unskilled ancient designer, or group of designers-builders who perhaps unconsciously included high levels of all five clues may produce a very thin and spiritual place. Yet, a very skilled and credentialed spiritual designer who unknowingly or purposely avoids including high levels of the five could intentionally produce a thick spiritual dud.

In a poem that follows, "Through the Gate" I recorded my experiences from my stumbling upon a Thin Place during an early morning walk in Kingsmill, Virginia, located between Williamsburg and Jamestown. That place is so history dense (clue 1) that the scene only needed an early morning quiet foggy river and a broken old gate to have that veil between worlds to become very thin momentarily for me.

That raises a question. Had I walked by the same scene at mid day, when the river was clearly illuminated, would I have been perceptive enough to feel the history density embedded in this area, where our country was born? Probably not. Thin Places will not necessarily be perceived as such 24-7 nor will they be perceived by everyone. As humans we have options in focusing our attention, setting our attitude and welcoming or refusing the invitation to spirituality.

I have found that poetry is one of the most accessible mediums to record Thin Place experiences, provided I take detailed notes very soon after the experience. If I try to revisit that Thin Place experience after many years, I find that prose works better. While some of my Thin Place experiences occurred over 55 years ago, they still continue to be so profound that I remember most of them as if they were yesterday. Several experiences in my teen years (written in prose) were not noted or otherwise recorded until many years later, yet they are still there embedded deeply in my memory.

Thin Places "talk" to us differently

Thin Places appear to "reach" us through transparent cells of one or more of these five clues, which attract alignments of two energy sources. One energy source is us, grounded here, on this side, along with the accompanying human strife, love, fears, faiths, anthropological patterns, and importantly, problems searching for solutions. The second energy source, on the other side, is Cosmic Truth, which includes all past history. Some may label encounters with Cosmic Truth as interactions with angelic messengers, or as part of the Will of the Universe, or simply God, or some other names. Whatever you call your Cosmic Truth and what I call mine is not important, particularly not to either of them (if in fact they are different). One of the most spiritual people I have known called her Cosmic Truth "Wow." I could not disagree with her, could you? And, she may have discovered that name at a Thin Place experience. Actually I did too, when I was 16, driving by an already well-known yet previously thick place for me before – and suddenly the sun-lit combination of what I know now as three of the five clues hit me. Even at 16, I had to pull the car over out of traffic as I just exclaimed "Wow!"

The brief essay entitled "Broad Street Station" is included later in this example of how Thin Places talk to us differently. They seem to get our attention through some form of intense beauty or expression of adoration. Then once we have been joggled into an openness, a wow (or maybe a huh?) may follow, then time shifts through generations or dense histories may follow. We often perceive silent messages that are difficult to put into words. We share the spiritual moment. Then it is gone.

Now, I ask you to reflect for a moment on a separate but related situation which you also may have experienced before. When I'm working on a project, maybe designing or writing a short essay or play, and I come to a point where I am yet undecided about what the next steps should be -- I often focus on it intensely, and purposely anchor a specific question or sought-after direction deep into my being, as if I'm placing it on a safe storage shelf within. Then I turn my attention to other activities such as mowing the lawn or engaging in some other mind-neutral activity.

Sometimes that may occur just before bedtime, and I'll go to sleep realizing that my subconscious may still be mulling over a problem at the point where I left off. Later out of the blue, I discover a creative answer to the puzzle or question I was carrying around. It might happen on the riding lawn mower, or during the night. I have in the past awakened at 3:00 am (seems like a favorite time) in the night laughing at a new play routine that just dropped in. A Thin Place in the night? Many people think so, and write accordingly, I was happy to learn.

My experience is similar when visiting a Thin Place. Within all of us, resides our human spirit, our soul, or our subconscious being to put it another way. Our human spirit has a special connection with Cosmic Truth, our "Wow," angelic messengers, or God, or the Universe's Will. We are hard-wired that way, much like the moth's obsession with a front porch light.

When we experience the presence of a Thin Place, we subconsciously perceive the stirring as a signal from Cosmic Truth, and that is when our human spirit puts those tingles up our spine. When we observe an awe-inspiring sunset with magnificent brilliant colors, or when we receive a "flash-answer" during the night, it is our human spirit that turns our antennae towards that invitation and locks in on it for a moment. During that lock-in, messages fly. In the process, we lose our sense of time and place. We receive impressions and messages based on our deep passion-based searching. I am reminded of the Bible verse, *"Ask, and it will be given to you; seek, and you will find; knock, and it will be opened to you."* Matthew 7:7.(NKJ) Receiving or being shown openings rarely occur on-target for me, unless I assertively prepare my asking, and passionately seek and continually knock. Then I positively expect an answer.

What I am suggesting here is that there appears to be a relationship between Thin Place <u>awareness</u> (and possibly its communications) and one's own personal passion-based asking, seeking and knocking for significant answers. Like a cell phone, communication is impossible if your phone is turned off.

During a Thin Place experience, we learn and grow. And once again, my friend, enabling such growth is the purpose of my writing this book, as I want to encourage you to seek out, nurture, visit and pause at the energy-charged transparent Thin Places. They are all around us, and easily found, because they usually find us instead. They may be disguised behind magnificent beautiful scenes, or sometimes they peek through mystical settings, particularly history-dense ones. They are there.

What does "science" say about this?

This question intrigues me, because science suggests not only the possibility of thin places (as not capitalized), but also the probability. Quantum physics, the physics of the sub-atomic world, has recently turned a number of conventional Newtonian laws of physics, which we currently live by, on its heels, disproving some of them and seriously questioning others.

For example, the scientific world has generally agreed with Albert Einstein's assertion that the speed of light is the universe's maximum speed limit. Nothing can travel faster. Yet now, quantum physicists have successfully shown that electrons can jump out of their orbits and relocate themselves vast distances away (essentially anywhere in the universe) and they do so instantly. In other words, they are not limited to the speed of light but relocate instantly.

Also, they leave no trace or track of their position in between the two points. They disappear from one place, and reappear instantly million of miles away. Both of these examples are contrary to conventional laws of physics, yet both are happening now. An excellent source of this "new science" information can be found in Leadership and the New Science, Discovering Order in a Chaotic World. Second Edition, by Margaret J. Wheatley, Berrett-Koehler Publications, San Francisco, 1999.

Another recent discovery from quantum physics is particularly relevant here and is also documented in Wheatley's book. Scientists can take paired electrons (similar to twins) and confirm that they are spinning in opposite directions. Then, if the rotating axis of one of the electrons is tilted, the other electron of the pair immediately adjusts its own axis to the same but opposite angle. Scientists have also shown that the distance between the pair has no effect on their ability to maintain a relationship and some form of communication. They can show, for example, that if we separate paired electrons and place one of them in an orbit around Jupiter and then tilt the axis of its spin, the other electron, still here, will instantly tilt its axis the exact same amount in the opposite direction. This astonishing connection is referred to as instantaneous non-local communication.

Now, how might this apply to Thin Places? Scientists know today there are about 23,000,000,000,000,000,000,000,000,000 electrons within each of us, depending on our weight. Consider the unlimited possibilities of communication pathways to and from us and other parts of the universe. Consider how many non-local communication possibilities there might be between worlds.

Based on the thin place evidence gathered and shared over thousands of years, coupled with the relatively recent discoveries in quantum physics, the number of non-local instantaneous communication possibilities before us is well beyond our human capacity to even count them or begin to fully understand.

But in similar ways to how we work with the invisible wind, magnetic fields and electricity, we also work with behaviors and reactions in order to partially understand unseen powers and harness some of the potentials for human good. We have the same opportunities with Thin Places: to learn as much as we can about how and why we are so influenced by certain configurations and links with history. Then, perhaps we can apply reasoning and structuring of Thin Place patterns and invest that knowledge in new environmental practices. That aim is what this book is all about. I begin in Chapter 2 with a recounting of what I experienced during my first Thin Place surprise when I was just 12 years old.

2

EARLY EXPERIENCES WITH THIN PLACES

It began at age 12

The Boy with the Kite

When I was about 12 years old, I was part of a group of neighborhood guys who gathered on a vacant lot to fly kites. On this particular day, the wind was just right. As we were piloting our kites, a well dressed young boy walked up to me and asked if he could join us. He had what looked like a brand new kite and a ball of clean white string. I had never seen him before. "Sure," I said," need any help?" He nodded yes. I handed my kite string to a friend for a moment. In a few minutes the new boy's kite was up there with the rest of ours. I got my kite string back and kept on flying. I looked over at the new boy, and saw him grinning from ear to ear.

Suddenly though, his kite began to do loop-de-loops. The boy froze, and his kite started looping faster, usually a sign of a crash coming up. But this was worse than a simple crash, because his was headed for the power lines. Several of us yelled, "Move back...pull it back!" Then we all watched it crash onto the top of the power lines. The little boy was horrified. He pulled on the string, but his kite had one of its corners caught under one of the wires, and it wouldn't budge. He dropped his string and turned to me with tears streaming down his face.

"Let me see what I can do," I said, realizing that this was basically hopeless. Then, I wondered if I could walk my kite over to his, pull my kite in some, and maybe nudge his just a bit. More than likely, this could be a two-kite fatality. My kite slowly got closer to his kite. I tried several times to lightly hit his kite, but each time, I would either hit the wires on miss everything. Then in one final attempt, my kite got stuck with his on top of the power lines. My suspicions of a two-kite fatality appeared to be confirmed. I had learned to never pull a tangled kite hard, because that just makes it worse. So, I stood there for a moment, among the silence of sadness, and waited – trying to figure out what to do. Then, a puff of wind blew by – and to my amazement it lifted my kite straight up and above the power lines. And, in the process, even more amazingly, it nudged the boy's kite loose. Cheers went up from all of us as we watched it drift to the ground.

The young boy ran over to the kite, picked it up and ran back to me, carrying his still fresh kite and now even a wider grin. When he got to me, I was ready to tell him, "You're welcome." No big deal. But when he got there, and we were facing each other about 3 feet apart, I saw the look in his eyes. It was powerful and penetrating, and I don't remember him saying any words at all, although I'm sure he did. His eyes revealed a very strong outpouring of gratitude and appreciation that I wasn't prepared for. I think I said, "You're welcome." At least I hope I did. Then the little boy was gone, with his kite. I never saw him again. Maybe he was from out of town visiting an uncle or someone near the vacant lot.

Over the years, I kept seeing that look in his eyes. I believe that's why I never forgot meeting him and his kite. In fact, his eyes haunted me off and on for many years. From time to time in my career as an architect, I would see that same look in someone's eyes I had assisted or helped in some way. Now, many years later, I wonder who was really looking through the boy's eyes out at me and leaving me with an important message. What a powerful delivery system! What a powerful message!

When I was 12 years old, of course I didn't know anything about Thin Places, or much else. I just saw a very strong expression from a person, and for some reason, I sensed I might learn more about this in years to come.

Later I learned that the Cosmic Truth has unlimited ways of reaching us with important messages. The history of Thin Places around the world suggests that such places may be one of the ways. And, now I believe that the small 3 foot space the young boy and I shared, at that very moment, became my first Thin Place experience. Little did I know that about 4 years later I would stumble upon another one.

Natural Bridge Virginia

There was a chill in the air one evening in June, 1950. I was a teenager attending a youth conference at Natural Bridge, Virginia. It was my first visit to the bridge, one of the seven natural wonders of the world, but I was skeptical about what could be so unique about this place. I understood it as essentially just a hole in the hillside.

It was near dusk. As we moved among the mountain setting through the ticket gates and along the paved pathway bordering a mountain stream, I could hear recorded music coming from up ahead, with crackles and scratches in the background, blaring from loud speakers, mounted high on the hillside. I still couldn't see the bridge.

Then, as we rounded a bend in the pathway – there it was. The voice on the speaker, blaring through the music, announced that it rose 215 feet tall — higher than a 15-story building — once owned by Thomas Jefferson, and surveyed by George Washington. Floodlights turned on and bathed the rock-faced mammoth natural sculpture in front of us. There was a huge arched opening in the mountain's rock face. Our eyes were pulled through the enormous archway along with the mountain stream and pathway, and beyond. The announcer faded and the music got louder, now featuring full choir honors. Even as a teenager, I was impressed. But I realized that much of this was theatrical. Meanwhile, the rock bridge stood in majestic pantomime, glowing under the light, as if wondering what all the fuss is about.

Then something strange began to happen. The recorded choir music continued. The group of us captured observers sat on benches along the pathway, gazing up at the massive structure, mumbling comments about the beauty. Goose bumps started running up my back. Maybe it was the chill in the air, I thought. I shuffled a bit as I pulled my old blue jacket collar up against my neck. It didn't help. Maybe it was the way the lights glistened off of the rock outcropping. Then, the bridge faded out of view a bit, and the scene appeared to change.

Suddenly, I began to sense something unexplainable. It was as if some mysterious secondary language was making itself known to me. Some undefined quality was being perceived through my young eyes. A sense of another level of meaning and expression started tickling my awareness here, nothing like I had experienced before. It was like hearing the beat of a song for the first time behind a familiar melody. I listened and watched more intensely. The glistening stone stood before us, steadfast and strong.

Then, I felt a radiation-like connection with what I was observing, an encompassing adoration and oneness accommodation to all of us below. I shifted in my seat again as unwelcomed tears washed my eyes and I wrestled to understand and push away these youthful embarrassing and unfamiliar reactions. Dave, a friend sitting next to me, sensed my uneasiness and looked over at me. "It's getting cold," I muttered back to him, shrugging my shoulders up towards my neck. "And the wind...makes my eyes water."

"Yeah..." He mumbled.

The feelings continued, and then subsided as the pageant program concluded. Later, we walked back along the pathway towards the entrance gate I wondered what had just happened here. What triggered these unusual perceptions and responses? All of us there experienced the feelings of awe from the massive size of the bridge, but there was something else that triggered my young antenna to tune into new and strange perceptions. There was a quality, a visual grammar, an unknown language that was beginning to poke through to me...and maybe to others.

What was it? Did the location of my seat impact my reactions? Are those reactions possibly predictable? I was full of questions. Now, about sixty years later, I still remember that experience like it was last week. Now, I realize this was my second Thin Place experience; this time introducing me to the visual language in architecture. But back then, I continued on my teenage trek pretty much on schedule – until about 2 years later driving by what was then known as Broad Street Station.

Broad Street Station

Later I had another visual ambush. Still a teenager and you can imagine I had other and higher priority issues to pursue than architecture. And, I almost succeeded in allowing those "Bridge happenings" to drift into a fading memory. But, at the age of seventeen, "it" happened again. I had another architectural wake-up call. This time, I was driving my first car, a prized black '41 Plymouth coupe, which I bought from a feisty retiring carpenter. I loved that old car. I was driving east on Grace Street in my hometown of Richmond, Virginia, one afternoon coming home from a downtown trip. As I approached the area of Julian's Restaurant, famous for magnificent pizzas, I turned over to Broad Street. To a seventeen-year-old, encapsulated in his '41 trophy and drawn primarily by appetite, I was in my own world of extraordinarily anticipated delight.

But, when I made the final turn, to head west on Broad, I was hit with another visual astonishment. This time it was Broad Street Station, the classic majestic old train station that I had seen at least a hundred times before. I had visited the building many times with my father. I usually played with the echoes bouncing from the marble floor and walls, and watched the frowned reactions of irritated travelers sitting in the vast lobby, surrounded by a garden of suit cases planted at their feet.

But on this day, the stately old structure smacked me across the face with its glowing prominence, and captivated me by its sparkling contrast among its industrial and commercial neighbors. As a teenager, I didn't understand any of this. But, on this day it grabbed my attention, blazing with a brilliant sun-lit resonance that shook me down to my grubby old tennis shoes. I gasped in surprise, and I still remember exclaiming "Wow!" Instantly, that Wow feeling flashed my memory back to the Bridge. It was the same feeling. I pulled over in a no-parking area and just stared at it. There was that radiation again. That undefined quality. That feeling of adoration just like at the Bridge.

Maybe it was the angle from which I viewed the building. Maybe it was the angle of the sun bouncing off the building form. It was as if waves of miniature signals were broadcasting from the train station's stone facade, through the shadows cast between the columns, and from the form of the dome against the dark trees and the brilliant blue sky in the background, all as if, again, it was a part of some type of visual language reaching out to me again. So there I sat bewildered in my idling Plymouth wondering what this was all about. What happened? Again? How? What was the basis for my reactions? How could I explain this experience to anyone?

Several weeks later, I changed my intended college major from art to architecture.

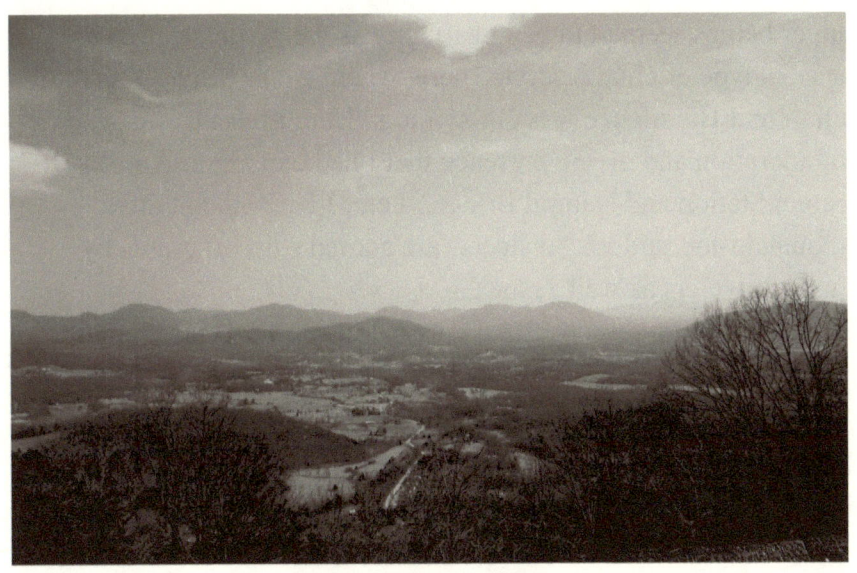

Afton Mountain

While studying architecture in college at Blacksburg, Virginia, I became very fond of living among the mountains. There are so many views which may have been further seedlings to my interest in what I know now as Thin Places. Yet it was closer to home near Waynesboro, Virginia, where I experienced one of the thinnest of mountain overlooks I had seen before. Back then, Route 250 was the main highway running up Afton Mountain, just east of Waynesboro. I remember the east side of the mountain very well with look-outs along 250, where people would pull off the road to look out over the vast valley space. One day I did just that on Afton Mountain.

As I parked my 1959 Volkswagen and walked out to the edge of the outlook, there was a noticeable quiet among the breezes blowing high above the valley. A small group of people were standing at the overlook peering out into a vast bowl of valley space. I glanced around, and even though I didn't know anyone else in the group, after a few minutes it seemed as though we all were brothers and sisters. While not a word was spoken, our inner beings seemed to be in harmony as we participated in a special type of church service here, where God's beauty was the choir and His silence was our sermon. Here I felt that same sense of adoration and divine presence that I had experienced at Broad Street Station and Natural Bridge. Later I learned that these mountain-top outlook "chapels" are adored stop-off points by many travelers as well as locals.

Later in my architectural career I began to wonder; are these mountainous Thin Place experiences nudged through the mountains masses, or through the huge blue mist-glistened cathedral-like space formed by the mountains (walls) and the valley (floor)? Perhaps both, but to me, it definitely is in the latter. Nature's architecture such as this is unparalleled in human connectedness. Its grandeur and intricacy in design, from miniature quantum particles to galaxies millions of light-years wide…continue to connect with us and always will. There are many such Thin Places in Virginia, throughout the USA and the world. I encourage you to Google "Thin Places" and visit some of the web sites of Thin Places around the world.

Now, let's consider the working details of those five clues which hover among and within Thin Places.

3

HOW THIN PLACES
CONNECT WITH US
What Powers these Super Engagements?

Backgrounds in Architecture

Many Thin Places are architectural in nature, such as monuments, towers, lighthouses, certain churches and more. Thus, Thin Places connect with observers through the same processes we use and refer to as normal architectural design and response criteria.   But Thin Places go beyond that and use connective tools rarely considered in normal architectural design. Of the five clues to Thin Place connectivity, three are used somewhat in conventional architectural design, some more than others, and two are seldom found as a primary design tool, except in what I classify as Thin Places. First, we will define those five clues (tools) and briefly describe how they work, so that you can see the picture as a whole. In the next chapter, we will begin exploring the five in detail.

The Five Clues

1 – History Density. When the Celts were driven from Europe to migrate in Scotland, Ireland, England, Northern Ireland and Wales around 500 B.C., they began building monuments and sacred sites supported by their pagan religions. After they converted to Christianity in the fifth century, they continued their long history of building towers, homes, monuments and gathering places; many of which no doubt sheltered them from oppressive enemies. Many Celts must have been killed in those structures and monuments, including entire families. Those monuments took on further "remembrance" functions for the Celts. Rituals using those monuments and places continued to develop. They would travel to certain monuments during religious celebrations, and then the journey itself became a ritual, known as pilgrimage.

Throughout many years, the sites harbored many memories of people, suffering, celebrations, deaths and more. When the Celts visited these sites, they were (and people still are today) overwhelmingly influenced by these history-dense structures and surroundings. Details on how this history-dense clue works is covered in the next chapter.

Following here is the first poem I wrote of a Thin Place experience, a **history dense** place where our nation was birthed. Early one November morning I was walking along the river's edge in Kingsmill, Virginia, getting a mile or so of exercise before our vacation breakfast. Mist was hovering over the James River -- that historic river just a stone's throw from Jamestown. The sunrise cast magnificent yellow tones up into the sky. Leaves were crackling, sliding along the road, and twirling in miniature slow motion hurricanes in that early morning off-season quiet. When I heard what sounded like voices echoing from the river, I stopped walking and gazed over at the river. An old gate was in my view, but I could see through the bars. A few moments later I had the same feeling as being on Afton Mountain, just a different seat with a river view. Slowly a poem began unfolding in my head.

Minutes later, I went back to the room. I told my wife, Kate I needed to get my camera fast and go back and get some pictures. I took a bunch of them, but by then the mist had all but dissipated. The sunrise had reached day-morning height. It was still a beautiful place, but now it had returned to a "thicker place," with people slamming car doors, and dogs barking in the background. I missed capturing the visual poetry.

I finished the poem after we returned home, and even though I had dozens of photos of the river, old gates, even ships at Jamestown, I never could get the correct visual interpretation of those Thin Place moments back in November. So the poem parked on my computer for about four years waiting for a visual partner.

Then, recently, as three friends were discussing the world's problems with Kate and me early one Sunday morning before church (at McDonald's), one mentioned a recent visit to Kingsmill and the beauty there – the same place Kate and I had visited. Unknown to our friend, Ilene, her words began painting the picture I was looking for. The following week I painted a small watercolor of what I remembered about the gate and the river beyond. I embedded the poem into a digital scan of the watercolor.

Through the Gate

Stale leaves crackled
against the rusty old gate
hung heavy on thick hinges
spiked deep in stubborn piers
near Jamestown, Virginia
hushed in off-season quiet
under a November dawn.

Its iron bars still gate tough,
with rods and hard welds
fixed like an opaque now, while
sensual spaces between bars
pave transparent avenues,
where dreams and time flirt
with the mooring river beyond,
among whispering ruffles, which
tickle tree roots along the shore.

And there, mingled in the mist
guarding its river secrets,
stands the veiled outline of a sailor
silhouetted below the yellow sky
staring across the river slick
at a small ship anchored in the fog
freighted with a handful of seeds,
navigated by gauges of passion,
fueled by winds yet forgotten.

He then leaned to the craft,
hollowed hands at his mouth
and megaphoned a discovery
out to tuned ears on board,
while I stood mute, peering
through the gate's cold grid,
stretching against today's bind.

"Bring the others!" He shouted
joyfully in English tone with
unquestionable authority,
lofted out at the shadowing ship,
berthed just off shore, and
a few hundred years away.

"And the tools,"
he hollered again, loudly,
bellowing - "This is the place!"

And so it was, soon to be,
when a grandson, now long dead
might have built this 2-way gate,
to divide and separate,
yet open a glimpse of history,
this place to birth a destiny
right here -- through the gate.

- - - - - - - -

We are all a part of a gate to potential Thin Places for others, aren't we? Thus, we encourage others to look not "at" us for all they will see is rust, broken pieces, bent hinges and other examples of the opaque now. We hope instead that we may inspire others to see the spaces between the bars -- the moments when a light shines through us, to illuminate the transparency of the Cosmic Truth, and His fingerprints all around and through us.

2 – Intrinsic Symbols. This clue-tool is also a very powerful one. Here the symbol has no man-made story; it doesn't "represent" anything but its influence is exerted through its intrinsic shape or presence and through universal reactions to it. An example is the color red. Unless that color has been "cultured' into a contrived story, we all will tend to respond to it the same. Light is also an intrinsic symbol, perhaps the strongest, which again all humanity will basically react to in the same manor. Obviously, one of the great benefits in frequent use of Intrinsic Symbols is that its compositions tend to hold human attention consistently. They tend to be timeless. Here is a simple little poem about a collection of Intrinsic Symbols.

Beauty

We are like the bridge,
which still shakes,
vibrating from its earthen ends,
spanning between logic and surrender,
rattling cold beams, slabs, cables
all substance to the core,

from the pageant, now passed,
with feet pounding like the big drums
and music swelling sweet purpled harmony.
Crescendo! Crescendo! Crescendo!
And we all resonated -- as One.

Even now after the procession,
the bridge still shakes,
for a while.

3 – Cultured Symbols. This clue is a very powerful design tool with a liability. A cultured symbol is defined as an object, sculpture, design form, pattern, plaque, even a style, etc., all tied to stories. Each symbol has a meaning intended for the observer but that meaning is anchored to the story behind it. When one looks at the symbol, he or she recalls the story and responds accordingly. For example a logo is a cultured symbol if it is so designed and if its story is properly maintained. There lies the liability and difficulty with this clue-tool. Stories behind symbols require high maintenance and usually the managers of conventionally designed facilities do not have the budget to properly maintain the stories' meanings among all users. Thus many of the symbols' communicative potentials die. One of the most popular and well maintained story-symbols is the Christian cross. On the following page is a story-poem that dances with some of those symbols.

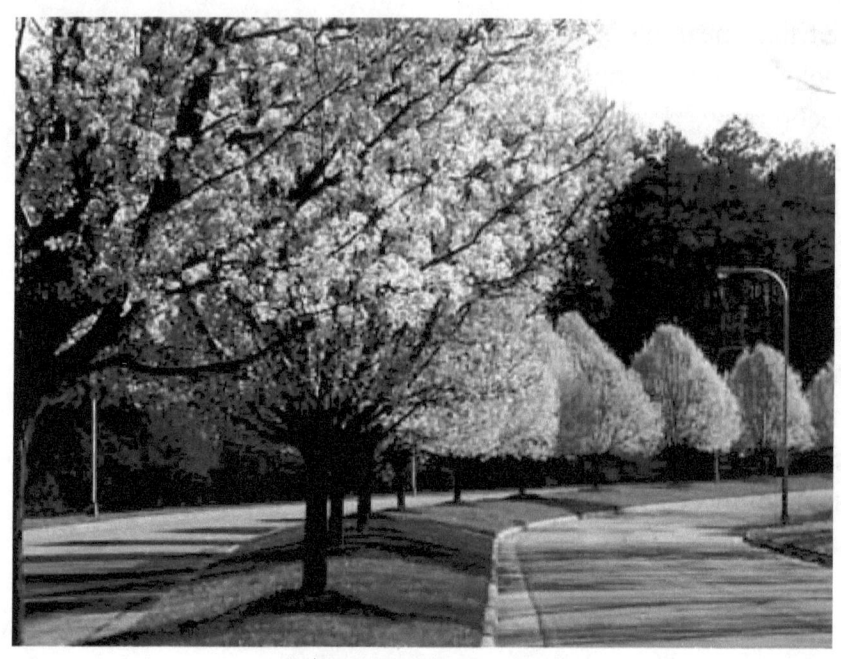

Spring Parade

Monks in white robes,
march through gardens of Lent,
in perfect measured step,
chanting awakening calls,
beckoning sacred ritual,
then they dissolve

into vanilla ice cream cones,
held high among the thicket,
celebrating joyful party toasts,
humming silent choruses, of
Happy Birthday dear Spring!
Then they transform again,

into glowing white fireballs,
hovering on charcoal sticks,
melting chilled March air and,
winter hearts, as another float
in Spring's parade, drifts by.

Now, after writing that poem, whenever I see white blooming Bradford Pear trees, I always think of monks, vanilla ice cream cones, and fireballs on charcoal sticks. But because that symbol-story is anchored within that particular poem, only those who read the poem, perhaps over and over, will make that same connection.

4 – Bio-Kinetics. Here the emphasis is on movement (Kinetics) and its influence on us (Bio). The primary focus is on one's movement, and attention focus as directed and orchestrated by the whole design experience. A design goal for any space is to arrange and design it in such a manner that its influence on users is complimentary with the purpose(s) of the space. This is expressed in Vitruvius's popular definition of architecture "Commodity, Firmness and Delight." For example, if a primary purpose of a particular space is social, but its design influences tend to be anti-social, it will be a problem space that will require high social maintenance, extra promotions, and multiple make-over renovations, all in attempts to compensate for the design's incompatible influences.

Movement influences are also produced as our bodies move as well. Obviously, it is important to get it right the first time. At Thin Places strong Bio-Kinetics are achieved with high intensity, much like high definition, and frequently at a spiritual level. Here is a short poem about kinetics and feelings. Notice the various movements of things, the bus, hands, and parents.

First School Bus Ride

It is the carrier of fear,
of dark imprisoned loneliness,
from young eyes lost,
stacked in rows,
peering through thick panes
gaping at fading rainbows,

as the bus drives off slowly,
while beckons of plastic smiles,
stretch from young faces inside,
and plead with the old, beyond,
through this rolling barrier's skin,
pulling away at the heart.

Still, it moves away slowly,
hearsing terrified souls
scholarly, with dry silent wails,
of abandoned desperation,
aimed at fading parents outside,
who slow-motion wave
empty mute strokes.

5 – Intention. This clue-tool is rarely understood as a design tool, yet it is one of the most powerful of the five. Consider intention in personal relationships. Most of us, after a few minutes in dialog with another person can begin to get an idea of the other person's intentions. If the other person is rude, arrogant and closed minded, we quickly get the reading. We respond accordingly, often in a protective and private way. Or, maybe the response is confrontational. Or, perhaps the response is a walk-away vacancy – no relationship. And a poor and inhibited relationship may seriously cripple any meaningful work the two might have produced together. On the other hand, if intentions are mutually compatible and empowering, so too is the tendency for an empowering relationship and heightened creative work product.

And so it is in architecture and particularly at Thin Places. A facility's intentions are conveyed quite purely through the design decisions. The design intentions should equal the unaltered expressions of the facility owner by conveying his/her facility intentions to the architect or designer. Unfortunately this is not always the case. At Thin Places, the intentions are unmistakably embedded deep into the place and they often have spiritual roots.

This leads to a couple of summarizing points applicable to all five of the clue-tools here as well as to architecture in general. It is of upmost importance for the facility sponsor(s) or owners to thoroughly discuss commit and convey the facility purposes and intentions to all parties involved. How else will those intentions get carried out if they're not absolutely made clear? This was one of the greatest successes of Disney's career – his conveyance of his intentions through fruition to his Disney World visitors.

Secondly, while using two or more of the clue-tools is admirable, as we will see in several examples of Thin Places, one of the clue-tools will usually dominate. Designing and managing those combinations will again require the skills of a qualified professional architect. Selecting cultured symbols, for example, and adding a story about the symbol which then requires maintenance of the story in order for the symbol-story to be understood must be done carefully. Suppose the symbol-story combination is intended to convey unity, and suppose the symbol selected to represent that with a story is a triangle – perhaps considering how the three sides of the triangle work in unity, etc. The problem here is that this is a mismatch of messages from the story and from the symbol. For example, should awareness of the story not be rigorously maintained, the symbol's meaning will be processed intrinsically (because no one remembers the story). But, a triangle is intrinsically processed more as representing power, bracing and rigidity than it does unity. The circle is an intrinsic representation of unity. So, in this triangle selection example, if the story maintenance breaks down, the messages intrinsically communicated may be contradictory to what was originally intended.

Here is a poem about a building type that has always fascinated me. Why are we so drawn to lighthouses? How can any of these five clues help us understand why we admire them so?

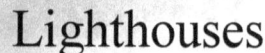

Lighthouses

What is it about you,
and your cousins alike,
that we so revere?

Is it your proud gantry,
mellowed, paint-clobbered masonry,
encircling, course by course,
stepping up, inching in,
honing the slant
like spreading feet
to brace your aim?

Is it your tall reach
which pokes through layers
of the blind mundane
to reach high above waves
ship-slicing rocks
wood-rot dampness
which you ignore?

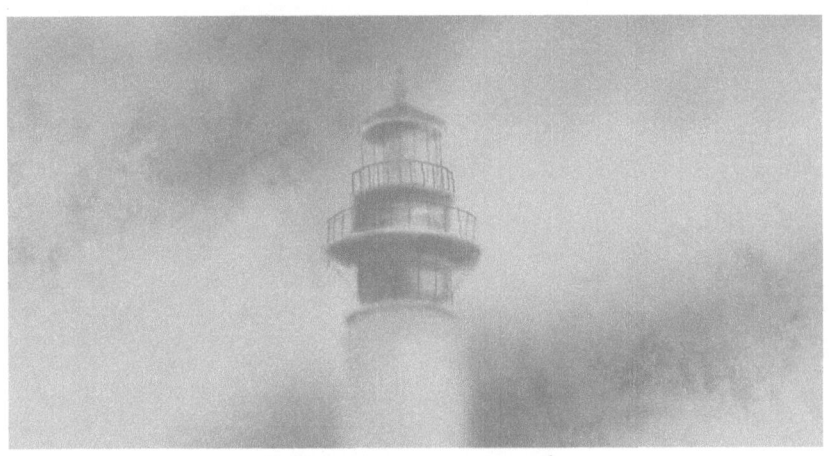

Is it your gaze outward,
your stare at the horizon
lifting us to the eye of your soul
the light, the optic nerve,
which paints as it sees,
peeks at distant seas
and buries lost images within?

Or is it the beauty
of your striped stalk
planted firmly with us,
against your dwarfed belly,
The keepers house
Of polished history,
Families and lens' cloths?

Why are we drawn
to gaze upon your stand?
Is it to honor your duty,
marvel at your form,
or rub beside and against
your noble intentions?

4

CLUE ONE
HISTORY DENSITY
The details of one of the ways
Thin Places are born

The level of History Density at a particular place appears to be a function of three elements: how many people might have been involved at this place before you; how long ago did the people involvement begin; and, what was the nature and intensity of the involvement? Let me illustrate with a simple yet profound example.

Have you ever noticed the clusters of flowers and religious symbols placed alongside highways marking the exact location where a family member or friend was fatally injured in an automobile accident? Does the marker say, "They died right here at this very place?" That's what I perceive. It is a sad reminder and my heart always goes out to those families and friends.

For a while, I wondered why the flowers were so precisely located, against a lamp post, or next to a scarred tree, or along a bridge. I wondered too, why not put the flowers at the gravesite or at the columbarium (if cremated)? I suspect if I asked that, the answer would be, "Because that's not where he died. He died right here!"

In my community several months ago, a two-year anniversary memorial service was held along a busy road for the death of a beloved policeman who was killed while he was attempting to arrest a motorist. The motorist started speeding away. The policeman hung onto the side of the car. The car flipped up onto an embankment and landed on top of the policeman. Clusters of flowers were placed there at that exact spot, and have been refreshed consistently for two years now. The memorial service was held at the same exact spot, because... "That's where he died!"

It is almost as if – when people die, especially like this, that at the very moment of death – when the human spirit leaves the body – that some of those (including some paired) 23,000,000,000,000,000,000,000,000,000 electrons, may have been left behind at that very spot, right there in the soil, or against the tree. Family members and friends can tell. Yes, he died right here! Something of him is still here. Have you ever wondered about that?

To the families and friends of those in this flower cluster example, they can intensely feel those three elements in this history-density example very strongly. It may have involved only one person, and maybe it was just last week, but the intensity of the involvement; violent death – was sufficient to mark this very spot here as a history-dense place, a Thin Place for them.

I wish now realizing this form of history-density in the making, that governing authorities in those areas in which death-marker flower clusters are placed, could instead of eventually chasing off continuous clustering at these spots, would instead come up with a safe and quiet marker design that the family could have permanently installed to honor their memory – and this new family Thin Place.

Now imagine the shock of neighbors and friends upon discovering that an entire family had been wiped out violently as they crouched terrified in their home in Ireland over 1,000 years ago at the hands of patrolling enemies. That place too may have been initially marked with flowers. Then more permanent markers might be erected in this now history dense place, and soon it may become a Thin Place.

There are other seeds of history density too. One day my oldest son asked me to go with him to visit a wood stove store. When we walked in, I noticed it was one of those "we have everything old here" stores. My son found the stove he wanted, and as he talked with the salesperson about delivery, etc., I wandered through the very ample supply of wood boxes, pots and furniture. I noticed a stand-up roll-top writing desk. I had seen stand-up desks before and roll-top desks but never as a combined unit or so I thought. Something drew me to the desk. I walked over to it, stood at the writing position, looked at the cubby holes for paper, pens, and the like, and suddenly, I had the strange feeling that I had had one of these before (although I knew I hadn't). I had very strange feelings standing at that desk. My son walked over and noticed me just standing at the desk which had grabbed my attention. "Why don't you get it, Dad?" I surprised myself by even hesitating, as I was actually considering buying the desk. Finally, I came to my senses and said, "No, I don't need it, there's no room for the computer, plus I can't imagine standing up all this time." (I was still convinced I actually had one at sometime in a past life, maybe). Perhaps some of you antique shoppers have had similar history-dense experiences.

Visiting places which are history-dense where original architecture, furnishings and accessories have been preserved is an ennobling way to "fish" for Thin Places. Businesses, universities, churches, families, and neighborhoods may all have very valuable histories, just waiting to reach out and empower us as well as to memorialize values. Instead of hiding those potential Thin Places only in museums, why not place those communicative jewels in and around the active pathways of people?

Framed photographs of relics with detailed and interesting text below describing how the relic was used, who used it, when, and how it may have impacted the development and growth of a company or university or church or family can then become a very powerful low-maintenance history-dense cultured symbol. Do you see the possibilities? Imagine the impact when these and other history-dense readable low-cost symbols are placed in office lobbies, libraries, dining facilities, and hallways.

Earlier in my career while chasing after these five clues, I felt the need to come up with some type of score sheet to help me more objectively evaluate places and compare one place with another. I started with listing 4 or 5 of the most reoccurring reactions and feelings I experienced when I observed each of the five clues. These would be guidelines listings only, because each place might suggest a totally new listing. But I would have the 4 or 5 previous reactions to start with.

For example, here are the five most popular reactions I have had when visiting history-dense places.

1 – I had a strong sense of history in the place.
2 – I felt a connection with me I couldn't explain.
3 – I wanted to stay here for a while.
4 – I wanted to be quiet and not talk with anyone.
5 – I had feelings of something very special here.

Next, I drew up a simple score box with rating ranges to record estimated values of each of these for me, for history density, at a specific site. Shown on the next page is how I logged my visit to the history-dense Virginia Plantation Chapel.

Example: Virginia Plantation Chapel

A = presence of item, 1 - 5 scale; **B** = weighted magnitude, 1-3 scale

1 - History Density	A	x B	= Score
Strong sense of history in this place	4	3	12
A connection with me I can't explain	3	2	6
Feelings that I want to stay here a while	4	2	8
Urgings to be quiet and not talk	5	3	15
Feelings of something very special here	4	2	8

Total score, this place, History Density = 49

Column A records the presence of each listed item on a 1-5 scale with 1 being a low presence and 5 being a high presence. Column B records a weighted magnitude, the intensity of presence of each item as I detected. The score then of each listed reaction is simply the value of column A multiplied by the value in column B. The total score for History-Density is the sum of all five reactions.

So, looking at the scores here, I was most influenced by the urgings to be quiet and not to talk, and secondly, by the strong sense of history in this place. I was least influenced by "a connection with me I can't explain," and secondly by "feelings of something very special here." There are, of course, no "right" or "wrong" answers. These are simply a record of my own personal impressions and reactions to a particular place within the context of its history density.

I developed listings and score sheets for each of the Five Clues and then hooked them together to have a Place Score Sheet that includes all of the Five Clues. That way, I (and you as well) can have not only individual clue scores, but also total place scores to compare the total "Engagement" score of one place with others. These will be illustrated throughout the book.

I set these up on simple spreadsheets with formulas to automatically do the totaling. You can do that too. Regarding the copyright, you have my permission to copy the blank forms at the end of the book as well as the forms with listings but no scores yet, for your personal use only (and not for resale), so that you can log and score your visits to special places. This will begin to help you clearly see those Five Clues, and perhaps become more attuned and aware of the presence of Thin Places.

While we have the Plantation Chapel example in our minds, I have included in the next pages a complete score sheet of all Five Clues and total score of this Thin Place. You will see my 5 reaction subjects listed under each Clue. The other four clues will be reviewed in detail in the next four chapters.

Example: Virginia Plantation Chapel

A = presence of item, 1 - 5 scale; **B** = weighted magnitude, 1-3 scale

1 - History Density

	A x	B	= Score
Strong sense of history in this place	4	3	12
A connection with me I can't explain	3	2	6
Feelings that I want to stay here a while	4	2	8
Urgings to be quiet and not talk	5	3	15
Feelings of something very special here	4	2	8

Total score, this place, History Density = 49

2 - Intrinsic Symbols

	A x	B	= Score
Space - identification	4	3	12
Light - present/contrasted	3	2	6
Proportions - intrinsic	4	2	8
Shapes - intrinsic	3	3	9
Color - intrinsic	4	2	8

Total score, this place, Intrinsic Symbols = 43

3 - Cultured Symbols

	A x	B	= Score
Stories anchored to this design	1	1	1
Stories behind forms and shapes	2	2	4
Stories behind layout	2	2	4
Stories - accessories & furnishings	1	1	1
Stories behind colors used	2	1	2

Total score, this place, Cultured Symbols = 12

4 - Bio-Kinetics

	A x	B	= Score
Motion - foreground, background	2	2	4
Motion in alignment	2	2	4
Motion between angles, curves	1	2	2
Motion as one moves around & through	2	3	6
Motion in transparency	2	2	4

Total score, this place, Bio-Kinetics = 20

5 - Intention

	A x	B	= Score
Of its being & purpose	5	3	15
Expressed in shapes-form-placement	4	2	8
Expressed color and honor	3	2	6
Expressed in Bio-Kinetic dialogs	2	3	6
Expressed in continuity	5	2	10

Total score, this place, Intention = 45

Virginia Plantation Chapel
Recap of Total Scores

	Tot		% of Tot
History Density	49	=	29%
Cultured Symbols	12	=	7%
Intrinsic Symbols	43	=	25%
Bio=Kinetics	20	=	12%
Intention	45	=	27%
Total Place Score, All =	**169**		

We have reviewed the five items listed under the History Dense Clue. We will review the items listed under the other four clues in the following chapters.

Looking at the Recap Summary above, we can see the highest impressions of this historic place. History Density ranked the highest among the Five Thin Place Clues, with Intention as the next highest. This is often the case (for me) in historic places. Intention usually rates high too. Intrinsic Symbols usually score in the top three clues in popular historic places.

You may see other patterns here as well. Remember, of course, that these are *my* impressions of a place you probably have not visited. Your impressions, if you did visit this plantation chapel, would no doubt be different than mine, at least in some areas. But I suspect many of the patterns and rankings would be very similar, unless one of us was previously conditioned (cultured) to respond differently than we normally would. For example, if earlier in life one of us was caught in a fire in a chapel that looked very much like this one; our responses would be quite different. Remember this point as we proceed, because this cultured conditioning process may sometimes hinder professional and best-choice decisions. Keep a team involved to help filter out off-the-wall conditioned responses.

As a final summary to History Density, following is a short poem of an old dairy barn which has been decommissioned.

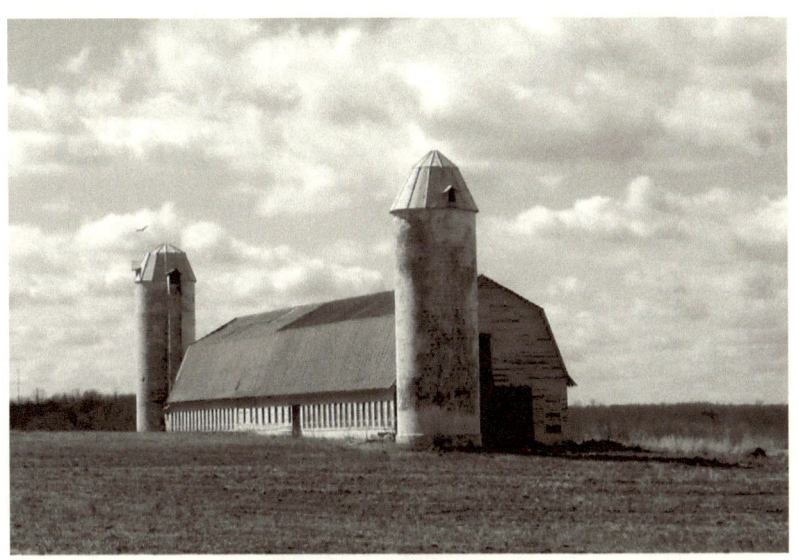

Sunset Barn

Warmed by the glow of today
shadowed in its own sunset,
the grand old barn sits barren
lingering on rotted feet,
worn stale of savored smells,

emptied of purpose, hollow,
moaning bone-drained abandon
of tomorrow's procession,
when dry seasoned land below
returns to voices young,

which skip among echoes
of prickly straw hats leaning
against stalled herds milked
long by enterprising knuckles
under rafters pitched proud.

Farewell -- old siloed friend
sustaining long to survive,
who waits now the tacit end,
when then, your final provide
will paint shadows in the wind.

5

CLUE TWO
INTRINSIC SYMBOLS

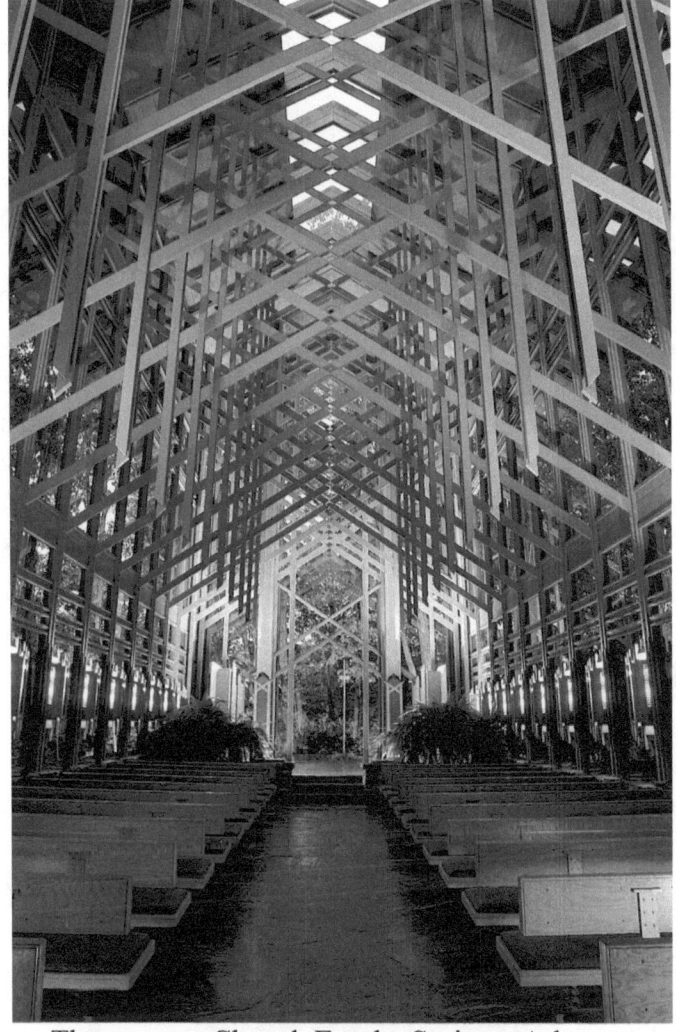

Thorncrown Chapel, Eureka Springs, Arkansas
Photo by Whit Slemmons

Symbols are those shapes, forms, emblems or other recognizable outlines which communicate meaning. They provide powerful ways to connect with us and deliver messages. Symbols make up two of the Five Clues in the architecture of Thin Places, the first being Intrinsic Symbols and the next being Cultured Symbols. Intrinsic symbols, such as light, space, and color trigger built-in responses which are basically the same for all of us, unless we are conditioned to respond otherwise. For example, we all respond to light pretty much the same way as the moth at the porch light. How we capture and use light within a composition raises or lowers the intensity of that response, but the fundamental response is the same.

Cultured Symbols, however, work differently. They are arbitrary forms anchored to a human-made story. We see the symbol, recall the story, and then we respond. A stop sign is a good example. We will explore Cultured Symbols in the next chapter.

Intrinsic Symbols include archetypes, primordial (ancient and historic) images, certain ratios, proportion, light, color, shape and other elements which have demonstrated over a period of time and across cultures as having similar universal and timeless influences upon the human psyche. Also included within a broad spectrum of Intrinsic Symbols are water, fire, earth and stone, air and wind, the moon, sun, the planets, other bodies in space, and plant and animal life.

In a sense, humans are hard-wired to respond to these symbols. More specifically, we respond to certain shapes, proportions, colors and symbols much the same way physical mass is affected by sound. For example, when an object with a given resonant frequency, is exposed to a sliding tone approaching a frequency which is the same as its own resonant frequency, the object begins to vibrate, echoing the same frequency. When the tone reaches the exact frequency as the natural resonant frequency of the object, the resonant vibrations are at their maximum.   The classic case of the crystal glass shattering under such resonant stress is a prime example.

In addition to our own resonant frequency in sound, we seemingly "vibrate involuntarily" when exposed to other stimuli intrinsically echoed in our own structure. Let's consider a few examples in more detail.

Space

Space is the cosmos's largest ingredient. Infinite emptiness. But to be perceived, it must be identified or defined. This can be done by making perceptible sub-limits obvious such as markers, walls, ceilings, etc. These then highlight smaller quantities of emptiness (space) between the observer and the next marker. As the whole space is identified, generally the more aware the observer is of the intrinsic archetypal emotion of "space".

A view into the Grand Canyon is an example. Observers standing at popular outlooks can be seen throughout the world staring into the huge voids for hours, feeling drawn to the great containers of emptiness and their identifiers, soaking in feelings of awe and various human rationalizations for the exposure-reactions to this intrinsic symbol of space. We observers, standing along the edge, are mesmerized by the "glow" of space.

Another interesting aspect of identifying space as an intrinsic communicator, are the tools which are used to sculpt the container.   Mountain ranges forming valleys, walls in great rooms, landscaping hedges, walks and lakes are all examples. Space can also be identified by tinting the emptiness with media. An example is fog or mist. Often a view across a plane or flat meadow which carries morning layers of sparkling droplets of moisture and mist are described in writings as having an unusual and moving "quality without a name." In such cases, the mist actually helps define the space. The same is true when we see billions of stars at night against the limitless backdrop of dark void. The "room" is visually defined as the great volume between the observer and the points (stars) which catch the eyes. This night volume is a huge Intrinsic Symbol. The distribution of points, brightness, color and other factors influence the dance of the observer's eyes partaking of the space. That dance is part of bio-kinetics, our Clue number four.

A helpful part in identifying space is understanding how our vision works with the brain to calculate and estimate distances for us. Since our two eyes are positioned in a horizontal plane, when our head is vertical we can see/locate a vertical line in the distance much easier than a horizontal line. As we attempt to focus on a point in space, our eyes scan the target and try to lock-in on an angle that each eyeball finds true at focus. Once our two eyes have established that lock-in angle, the angle is sent to the brain so it can use its built-in trigonometry to calculate the distance. When we look at the vertical edge of a cabinet across the room, for example, and then say, "Well, it's about 12 feet or so away," that's the process our eyes and brain go through in a few milliseconds to give us that information. Now try looking at and focusing <u>only</u> on a horizontal edge of the cabinet See how much more difficult that is? Our eyes seek vertical lines to help identify distances, and thus process the perception of space. When we can identify distances and thus perceive space, our intrinsic receptors vibrate. Perceiving space is exciting to our psychic. And, vertical lines enhance that perception.

As described in <u>Psychology, Themes and Variations</u>, 5th edition by Wayne Weiten, Santa Clara University, published by Thompson Learning Inc., 2001, David Hubel and Torsten Wiesel, both scientists, also found vertical lines to be "exciting" to a cat they had wired into their microelectrode testing devices. When the cat was exposed to a vertical line, the screen recorded 14 firings of a neuron in the cat's visual cortex section of the brain. A horizontal line produced only two firings, and a diagonal line produced five firings.

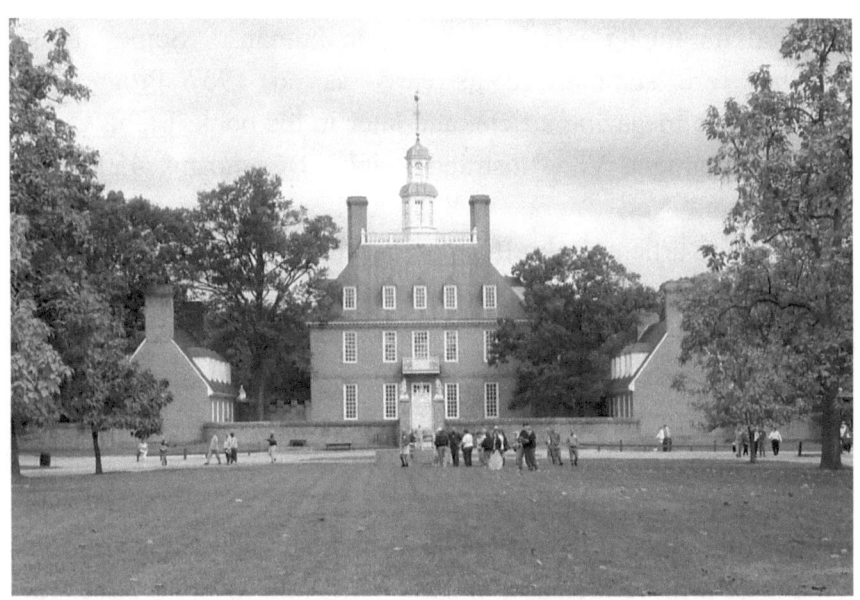

Shown above is the historic Governor's Palace in the restored Colonial Williamsburg, Virginia. Note the large open green space in front of the Palace, identified by the trees on each side, the roadway and the Palace wall. Then, notice that the Palace is "looking at" this space, a subliminal yet engaging acknowledgment for observers. Then, look at the space just inside the Palace's front wall. Notice how this space is identified just enough so that we can respond to its intrinsic symbolism. See the two buildings, one flanked on the left and one on the right. See how your eyes move from one, across the front garden wall, to the other, then across to the Palace façade and the trees. The space is defined – and the space itself is like a transparent building. Can you see it there? Like a small mountain valley, it is there. Identified emptiness is more than nothingness. And our feelings toward it are comfortable, intrinsic and consistent.

Color

Color is an intrinsic symbol. Although we all may have color preferences, usually through cultured preferences we actually all respond to color the same way. Faber Birren, author and color consultant, has done much research and published significant material on the effects of color upon humans. Some of his findings, are summarized as early as an 1967 Progressive Architecture magazine article, and later in his book, Light, Color and Environment, Van Nostrand Reinhold Company, 450 West 33rd Street, New York, N.Y. 10001. In the Progressive Architecture article, Color It Color, September 1967, page 131, he said, "Bright light and warmth of color tend to condition the body for physical action (like a sunny day), whereas subdued light and cool color (like a cloudy or rainy day) are more conducive to moody reflection."

Further in the same article Birren states that, "Optical stimulation through brilliance of light and warmth of color will cause a number of things to happen – and without human volition. First there will be increased muscular tension. Second there will be attraction to stimulus. That is, the outstretched arms and the body itself will tend to lead toward the bright light and vivid color. Third…respiration will increase, so will heart action, while blood pressure will rise. Fourth, there will be increased cortical (brain) activity, which can be electronically recorded. Dim light and cool color will tend to have the reverse effects. All or most of this will take place independent of any mental opinions."

Tendencies to respond towards predicable reactions are also found in Jung's writings of archetypes, described originally by Jung simply as primordial images, as noted in "C.G. Jung, Word and Image" edited by Aniela Jaffe, Princeton University Press, 1979. Jaffe quotes Jung as saying (the archetype) "…is an irrepresentable (*his word*) unconscious, pre-existent form that seems to be part of the inherited structure of the psyche and therefore manifests itself spontaneously anywhere, at any time. The representations themselves are not inherited, only the forms, and in that respect they correspond in every way to the instincts, which are also determined in form only."

Shapes

The circle is a primordial image, an archetypal form, and perhaps the strongest of all architectural Intrinsic Shape Symbols which trigger universal human instincts of wholeness and oneness. For the same reasons and background, I would place the square, the spiral (based on the Fibonacci series), the equilateral triangle, and the rectangle (based on the Golden Section) in the same intrinsic library. As we shall see later, these particular intrinsic forms also have quite strong bio-kinetic influences as well. It is interesting to note that the rectangle based on the Golden Section demonstrates a relationship between sides (proportion) which have been shown to have universal appeal. Note, it is the relationship, the vibration of two expressions, which appears to be a key factor here, not the object itself.

The Golden Section

In fact, it is the Golden Section relational vibrations (proportion) which are like resonant frequencies to our human nature. It is not the math of the proportion, but a more rooted and simpler premise of the section which is noteworthy. Summarizing its basis: if we were to divide a line into two parts, there is only one point (and its mirrored counterpart) at which the relationship of the smaller part (section) to the larger part is exactly the same relationship as the larger part to the entire length of the line (the whole). If you assign a unit of one to the length of the line, the Golden point is at approximately .618... from the end. Much has been documented on the extent of nature's dimensioned parts, including humans, being made up of these same relationships. The Power of Limits by Gyorgy Doczi, published by Shambhala Publications, Inc., 1994, illustrates dozens and dozens of uses of this Golden Section proportion in nature, art and architecture. People simply tend to be comfortable with built relationships of the same proportions, perhaps because our own body dimensions use harmonics of the same proportion. Architectural examples of the Golden Section go back to include Stonehenge, the Pyramids and even further back in time than its own definition. Ancient designers recognized this proportion and its continued use supports its intrinsic nature. The relationship, the vibration between the two dimensions, is the Intrinsic Symbol.

Intrinsic Symbols and Duration of Influence

If all other contents are set as equal between two example environments, except Intrinsic Symbols, -- then, the length of time in visual connectedness of the environments, and the positive visual impact upon observers and participants in each environment, will be found to be in proportion to the quantity of Intrinsic Symbols embedded into the design. In short, the timelessness of a facility or landscape design is a function of many aspects, including design talent and the skilled use of the other four clues here. But, one of the other high factors is the abundant presence of Intrinsic Symbols.

Score Sheet response subjects for Intrinsic Symbols

Referring back to the Thorncrown Chapel, I used my Score Sheet to measure my responses to the Intrinsic Symbols here. This is what I found to be important to me as I looked for Intrinsic Symbols here. Yours may be different.

Space – identification: Look at the vertical lines in this interior space. And look at the transparency of the clusters of vertical lines (truss members) as they align in the foreground and background. Space is excitingly identified. This is one of the reasons I gave this Thorncrown Chapel a perfect score of 75 on the Intrinsic Symbols clue.

Light – presence/contrasted: Look at the light filtering through the space here. Very beautiful.

Proportions – intrinsic: A laced multiplicity of rich proportions.

Shapes – intrinsic: I see many shapes, partial shapes and lines from nature here. Do you see the shape of a partial snow flake, as the room reaches upward in praise? Do you see the diagonal lines replicating the diagonal rays of light we see even from the stars?

Color – intrinsic: Difficult to see in this grayscale photo, but the colors here are all earth tones. Another outstanding feature of this place is seeing and connecting with the outdoors from just about anywhere inside.

You might notice that I scored Thorncrown very low on the Cultured Symbols clue. That is because there are very very few. That may also reflect one of the reasons that this place received another AIA award, this one for being one of the best 25 year old facilities in the USA. Intrinsic Symbols are timeless. Cultured Symbol's life is dependent upon its story maintenance. The fact that Thorncrown contains very few Cultured Symbols in its design was not an accident, nor an oversight. It was an intentional wise and skilled decision by its architect, Faye Jones, FAIA. You will find the same choice tendency at many Thin Places. When Cultured Symbols are used, they are very often part of the decorative detailing, a part of the textured elements used in the overall composition. This provides options for observers without libeling design engagement. For example if the observers wish, they may pause and reminisce through the Cultured Symbol stories embodied within the textured detailing – but if they do not so wish, or if one cannot remember the stories, the decorative texture simply continues to work very effectively through the Bio-Kinetic clue – pure chorography of eye movement.

Place: **Thorncrown Chapel**

Eureka Springs, Arkansas

1 - History Dense	A	x B	= Score
Sense of long history here	4	3	12
A connection with me I can't explain	5	3	15
Feelings of wanting to stay a while	5	3	15
Urgings to be quiet and not talk	5	3	15
Feelings of something special here	5	3	15
Total score, History Density =			**72**

2 - Intrinsic Symbols	A	x B	= Score
Space - identification	5	3	15
Light - presence/contrasted	5	3	15
Proportions - intrinsic	5	3	15
Shapes - intrinsic	5	3	15
Color - intrinsic	5	3	15
Total score, Intrinsic Symbols =			**75**

3 - Cultured Symbols	A	x B	= Score
Stories anchored to this design	2	1	2
Stories behind forms and shapes	2	1	2
Stories behind layout	2	1	2
Stories - accessories & furnishings	2	1	2
Stories - colors used	2	1	2
Total score, Cultured Symbols =			**10**

4 - Bio-Kinetics	A	x B	= Score
Motion-foreground, background	5	3	15
Motion in Alignment	5	3	15
Motion between angles, curves	5	3	15
Motion as one moves around-through	5	3	15
Motion in transparency	5	3	15
Total score, Bio-Kinetics =			**75**

5 - Intention	A	x B	= Score
Feelings of purposes beyond utility	5	3	15
Expressed in shapes-form-placement	5	3	15
Expressed in color and honor	5	3	15
Expressed in Bio-Kinetic dialogs	5	3	15
Feelings of honor and adoration	5	3	15
Total score, Intention =			**75**

Place: **Thorncrown Chapel**
         Eureka Springs, Arkansas

A = Presence of this element in the composition, 1-5 scale

B = Quality Value or Story Maintenance on a 0-3 scale

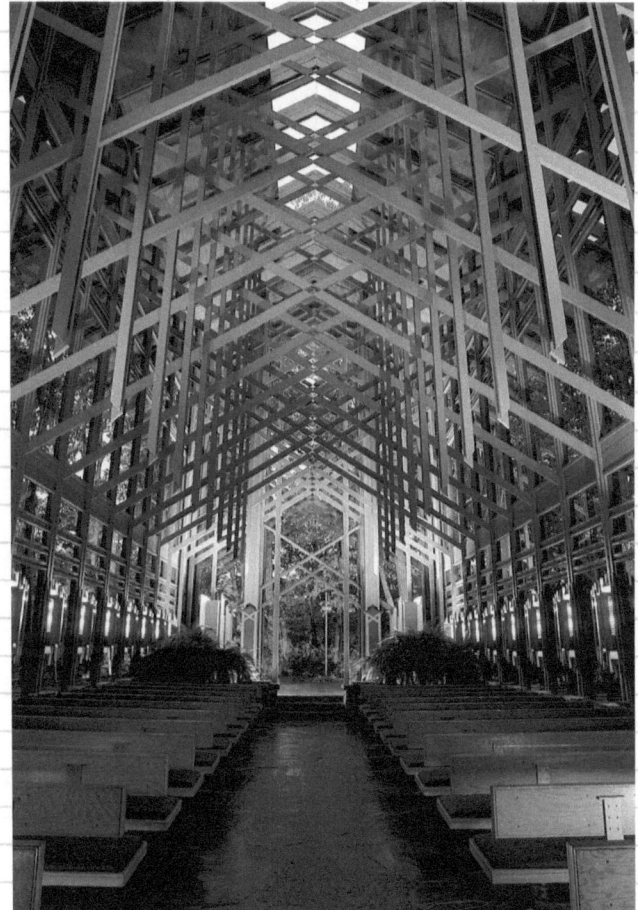

Photo by Whit Slemmons

Recap of Five Clues Scores

Total score, History Density =	72	= 23%	of Total Score
Total score, Intrinsic Symbols =	75	= 24%	of Total Score
Total score, Cultured Symbols =	10	= 3%	of Total Score
Total score, Bio-Kinetics =	75	= 24%	of Total Score
Total score, Intention =	75	= 24%	of Total Score
Total Score, All =	**307**	(of possible total 375)	

6

CLUE THREE
CULTURED SYMBOLS
Like fashion, the end results
will change with the stories

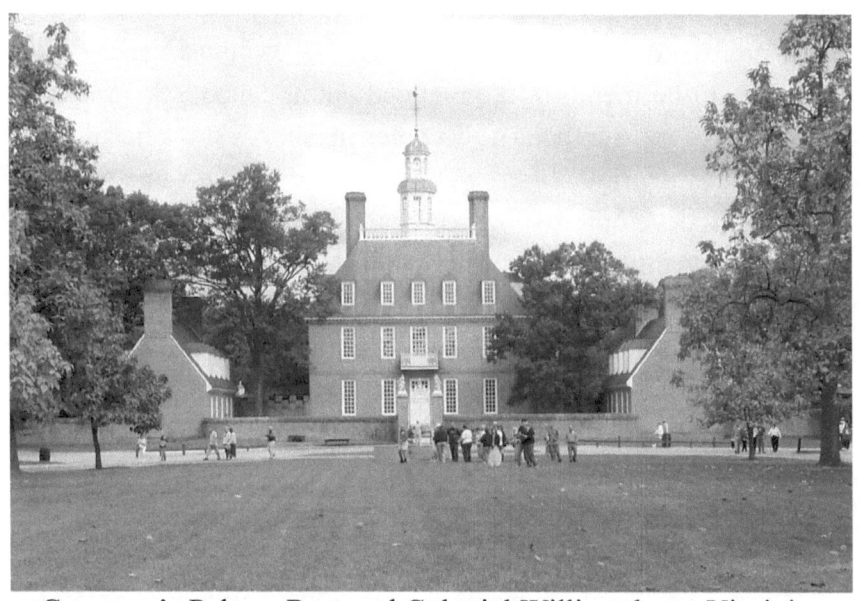

Governor's Palace, Restored Colonial Williamsburg, Virginia

Cultured Symbols differ from Intrinsic Symbols primarily in how their symbolic meanings are remembered. We learned in the previous chapter that Intrinsic Symbols' meanings are "hard-wired" into each of us. The meanings are intrinsic. <u>Cultured Symbols</u>' meanings, however, are based on <u>stories</u> anchored to an object (the symbol). Imagine as an example someone telling us a story, then holding up a vase and illustrating the story by associating the vase's design and history with the story. If this story association is performed over and over again, fairly soon, when we are shown the vase – we will recall the same story. That is, as long as we can remember the story. The vase is now a Cultured Symbol.

An example of a well-maintained symbol story

A popular Cultured Symbol is the Christian cross...a widely used symbol story based on the crucifixion of Jesus. Another example of Cultured Symbols is the alphabet. Each letter is memorized, with its story of sounds and shape. The story of each letter is reinforced during school and reading exercises. Then there are word stories, again learned as symbols, each with meanings to be memorized, practiced and recalled. Soon, when we observe the symbol, it becomes a short-cut to recall the meaning of the original (letter or word) story. The same process is used in developing, observing and meaning recall through architectural cultured symbols.

Cultured architectural symbols may be simple pictures, statues, plaques, emblems, or design forms crafted to express and recall meaning or association with an original story. The short-cut process is powerful and works consistently as long as the story and symbol connection is consistently maintained through re-telling of the story-symbol connection over and over so that the connection endures throughout generations.

Intrinsic Symbols as we have seen in the previous chapter do not work that way. Another way to understand the distinction between Intrinsic Symbols and Cultured Symbols is through their memory sources. Intrinsic Symbols utilize primordial memory – memory with which we were born – and which acts much like human instinct. It's similar to your computer's ROM (read only memory) which is built into the computer. In contrast, Cultured Symbols use a shorter term, writeable (and forgettable) memory; similar to letters or spreadsheets you create and then save on the computer hard drive. When we retell from memory the story of Little Red Riding Hood to a group of youngsters, we are using that shorter term memory to recall the story the best we can.

"Looks like" is rooted in short term memory

Now, back to the vase example for a moment – if I were to hold up that same vase in front of a new group of people and ask what it means to them, there would be understandably be a variety of responses. Each person would scan his/her associative symbol libraries for memories of an object similar to this vase to recall any meaning. One person might say, "Yes, that looks very much like the vase my grandmother handed down to my mother...and which I broke into a thousand pieces when I was four years old. I will never forget that experience."

"Looks like" is a key that one is speaking from the shorter term writeable and forgettable memory. Some people in the group may identify exactly the age and origin of the vase and the story behind its design. This is still from our short-term memory. Others in the group may not retrieve any meaning from their associative memory scan of this particular vase. They might respond through reactions to the intrinsic sculptural form of the vase or through another routing entirely which we will explore in the next chapter as Clue Number Four, Bio-Kinetics.

Now, many examples of Cultured Symbols in architecture are found among religious symbols. Consider the fish symbol. What does that symbol mean to you? The original meaning goes back to early Christian persecution when they found it necessary to use secret markings and drawings to signify to other Christians that they were in fact also a Christian. The signaler within a small group would often draw in the sand a simple outline of a fish. Other Christians in the group would immediately know that person as a Christian because they knew the secret symbol. That is a simplified version of the story. When you see that symbol as a bumper sticker on the car in front of you – now, you may recall the story and process its meaning accordingly. You may also know that this Cultured Symbol is still working, at least for you and the person driving, for now.

Two keys are important here in considering the use of cultured symbols and finding them in Thin Places. One is the story, the meaning behind the symbol. The second key is the maintenance of the story. Like the alphabet, the symbol-stories must be maintained over and over to new generations to reinforce and maintain the association. Consider how many times each day we do this with the alphabet. Without maintenance, the symbol will die, as C.G. Jung describes in his writings.

And, therein lays the huge liability in the use of Cultured Symbols within Thin Places or other architectural examples. Without adequate story maintenance, they become dead and thus simply don't communicate to observers what was intended. Some die very quickly.

An example of a dead symbol

Here is a real-life example. Years ago I was asked to comment on the potential renovation needs of a relatively young (1970's) East Coast college student activities building which is a part of the original historic campus dating back to the mid 1800's. I was told that physically, the facility was not in bad shape, but functionally; it had some serious problems although it had been originally designed to serve as a student center. Upon visiting the campus, I learned that when the building was dedicated, the donor who had financed the facility arrived from California at the local airport for the special event. When a college official drove the donor to the campus and alongside the ribbon flying facility, the donor exclaimed, "Is THAT the building?"

When the answer came back as "Yes," the donor asked immediately to be returned to the airport. Student and staff reactions have been about the same since it was built. The design of the building was an echo of a popular 1970's "brick masses with ribbons of windows trendy motif." The style was a cultured symbol of the time, plus it was of an "imported culture" from someone's far away drafting room or magazine rack which had no real visual connection with the site here, no connection to the original historic campus, or local edification, nor even with student center user needs. Students and faculty were left with the feeling that the building design might be a symbol of the future, but it was speaking another language, and what it was saying, apparently, had little to do with a "student center." And, they were right.

Engagement failure

The reasons behind this building's engagement failure were actually three-fold. One, we humans are addicted to symbolizing nearly everything from cars, sports, language, fashion and even people. We take the short cut. We symbolize. Symbolizing is also the root of prejudice. So here, a previous original facility design (probably an award winner) with certain inventive features was symbolized into even fewer features. A symbol, a short cut description, is more easily communicated and replicated than all of the original details and design wholeness.

Two, the building symbol story was never anchored to the place and people to which it was to serve. Three, the unacceptable design intentions (clue 5) telegraphed through so strongly that it revealed a lack of place-integrity from its very soul. Observers and occupants felt a sense of hypocrisy and betrayal in its midst. No one enjoyed being there. Have you ever visited places like this?

This building and hundreds like it were designed based on symbols (as design concepts) cultured within architectural magazines and design departments across the country. To the designers, this approach was a remarkable confirmation of what was honored at the time within the "industry" as THE NEW APPROACH to good design. Even the dark dingy brick-lined toilet stalls reflected such marvelous continuity. The trouble was that no one got it, except perhaps the small groups who attempted to perpetuate the trend. The cultured design symbols used here were dead before the donor's airplane landed. The college agreed. What a horrible waste.

How can we help prevent these disasters? Would preliminary discussions of an aesthetic model (similar to this Five Clue review) have provided some objective basis for those charged with the responsibility and opportunity to informatively guide the programming and development of this facility? For example, what if the building committee and the Board of Trustees of this prominent college had even an elementary understanding of this Five Clue model? What kind of dialog might have developed if the Board of Trustees had insisted that the architects explain in their presentations how their proposed design utilized these five clues (or had used another similar model) which would spring from the root of the design?

Instead, what happened here was a non-representative sub-culture essentially designed the building. Unfortunately, this continues to happen in architecture today. There are mutated processes which in reality are not advancing the art of architecture nor are fabricating environments which enhance and ennoble the human experience. Instead, many are dedicated to enhancing the replication of a stylized and symbolized processes discovered or contrived along the way.

Study designs which highly engage

Are building owners, developers and designers failing to understand physical exposures and formative relationships which cause certain and predictable human responses? Have they visited Thin Places? Are we failing to advance our art? And, because of this possible failure, do standard design procedures often fail to produce facilities which optimize their connection with the human spirit of users and occupants. In the process, are we lessening the value of the human experience with architecture and the physical environment? And, unfortunately, with each new crude and baseless example of human-less streetscape that is added to the scene, it shouts a lesser value of the individual.

Cultured Symbols can be powerful as well as ruinous in environments, and their use must be applied very carefully. The success of the process tends to be proportional to the understanding and acceptance of the story behind the symbol, and then maintaining the awareness of that story. Often such meanings of a symbol are positively reinforced and maintained through ongoing educational, religious and corporate systems, and thus, their human connective competency continues strongly. However, there are many examples of misuse of this design tool where adequate maintenance is not feasible, nor even worthy. Thus, the story never connects, or fades, and the symbolic premise dies, leaving a barren, oblivious form which bio-kinetically may radiate despair, failure and gloom.

As we know, trends come and go, and probably will do so always. While there may be merited parts and pleasant refreshment in a new trend, fashionable architecture is a "timed way of building" contrasting that with the preferred timeless way. I don't mind throwing out an old leisure suit (remember those?) but it is disturbing to see 30 year mortgages financing three year trendy symbols. Style itself is intrinsic, but a particular style is cultured and, thus, its long term connectivity may be whimsical.

Symbol maintenance is expensive

Unfortunately, most sponsors of facilities are unwilling or don't recognize the need or cost to maintain symbol-story relevancy. So, many of the good intended story symbols soon die. We may still derive some visual pleasure from them but that lifeblood is usually through one of the other four response clues and not from a forgotten story.

This leads to another danger of cultured symbols. Cultured Symbols are often designed to represent a story or event, such as crossed swords representing some form of badge of two forces, meeting or engaging, or merging...depending on the original story. Yet, if the story dies, we then digest the design through other modes. The difficulty arises when the design communicates visually, through bio-kinetics for example, and does so with a contradictory or negative response. An example I frequently see is the "X" repeated in church facilities, in woodwork, grilles, screens and other architectural detail. I suspect that most observers do not make the connection between the X and the Crusades, or other X story sources. Yet, the X is bio-kinetically disruptive to the eye path, particularly in clusters, as in a grille design. What started out as an intention to symbolize and recall Christian heritage, for example, may now be misunderstood and instead perceived subconsciously in its pure visual sense as a disruptive impulsive brace, a do-not-enter warning. A circle would be a much better form to use in the grille design.

The golden section and its harmonics are important design tools, as the circle, square, hypotenuse of a triangle and other relational compositions. Yet, they are only tools. Most any talented designer, as an exercise, could use ALL of these and any other recognized geometric forms or prescribed ratios, or other mathematical rules to create (on purpose, again as an exercise) the most awful facility one might ever experience. These tools are notes in a symphony...or noise. The composition is a function of many influences which experienced architects understand..

Corporate logos continue to have meaning as long as the corporation chooses. The Christian cross will live on, perhaps forever, as will other religious symbols which are reinforced by the repeated telling of the story behind the symbol. College mascots, logos and facility displays representing the history and significant events of the institution are a few other examples.

Sacred landscapes are rich in the awareness and knowledge of representative symbols.

Cultured Symbols telling stories of a place's History Density and Intentions

I urge building owners and users to intentionally participate in digging up old symbols, pictures, objects, columns or whatever else is part of its own history density and consider using these cultured symbols to communicate their story and intentions (clue 5). The building team can go over the collection very carefully and work selected ones into the voice of the facility, reinforced with text as necessary to recapture and re-tell the story.

During the design of another student center, for example, a college committee was formed to dig and gather such significant memorabilia. Then, a design staff person worked hours and hours with the committee to flush out the best collection and prepare necessary text to relay the story, the results of which now are woven throughout the spaces within the center. As I would visit the center often, I would see grins on the faces of fresh young students as they lean towards wall plaques to read text and photos, absorbing the committee's work, the history and antics of their peers, how the historic dorm was first built, what the bell tower story really means, and why the goal posts were put in the lake – all while having lunch with two friends and a new professor. It's a joy to see the process working. New stories and representative symbols are studied through the committee and text added as necessary, then judged as expressions of the college's culture. Similar processes can be utilized to activate the power and pleasure of cultured symbols for all facilities and their people relationships. This is important.

Cultured Symbols and
Score Sheet response subjects

Stories anchored to the design: I was trying to measure and log reactions here at the Governor's Palace in the restored Colonial Williamsburg, Virginia. In this example, I was looking for Cultured Symbols in the design, and if I perceived any, yet didn't know the meaning or story, I scored that low. For example, if I saw a stop sign, I would know the meaning without having a written description of it on the wall. Our day-to-day experiences do a reasonable job in maintaining the story behind that Cultured Symbol. But if I saw a yellow sign with the word "Endeavor" written on it, I would be at a lost without some type of story to help me here. Without it, I would score this observation low. Note that I found a fairly substantial number (column A) of Cultured Symbols, but usually did not know the story behind the symbol. So, sometime between when it was built and now, the symbol has died (for me).

Stories behind forms and shapes: Found some, but stories were unknown and unsupported as far as I could tell.

Stories behind layout: fewer found, but no story.

Stories – accessories & furnishings: More found here, and a few plaques and casework text helped.

Stories behind colors used: Very few found.

Place: **Governor's Palace, Restored Colonial Williamsburg, Virginia**

1 - History Density	A	x B	= Score
Sense of long history here	5	3	15
A connection with me I can't explain	5	3	15
Feelings of wanting to stay a while	5	3	15
Urgings to be quiet and not talk	5	3	15
Feelings of something special here	5	3	15

Total score, History Density = **75**

2 - Intrinsic Symbols	A	x B	= Score
Space - identification	5	3	15
Light - presence/contrasted	4	3	12
Proportions - intrinsic	5	2	10
Shapes - intrinsic	5	2	10
Color - intrinsic	5	3	15

Total score, Intrinsic Symbols = **62**

3 - Cultured Symbols	A	x B	= Score
Stories anchored to this design	4	1	4
Stories behind forms and shapes	4	1	4
Stories behind layout	2	1	2
Stories - accessories & furnishings	5	2	10
Stories - colors used	2	1	2

Total score, Cultured Symbols = **22**

4 - Bio-Kinetics	A	x B	= Score
Motion-foreground, background	5	3	15
Motion in Alignment	5	3	15
Motion between angles, curves	5	3	15
Motion as one moves around-through	4	3	12
Motion in transparency	3	2	6

Total score, Bio-Kinetics = **63**

5 - Intention	A	x B	= Score
Feelings of purposes beyond utility	3	3	9
Expressed in shapes-form-placement	4	3	12
Expressed in color and honor	4	3	12
Expressed in Bio-Kinetic dialogs	5	3	15
Feelings of honor and adoration	4	2	8

Total score, Intention = **56**

Thin Places and Five Clues - Matrix

Place: **Governor's Palace, Restored Colonial**
Williamsburg, Virginia

A = Presence of this element in the composition, 1-5 scale
B = Quality Value or Story Maintenance on a 0-3 scale

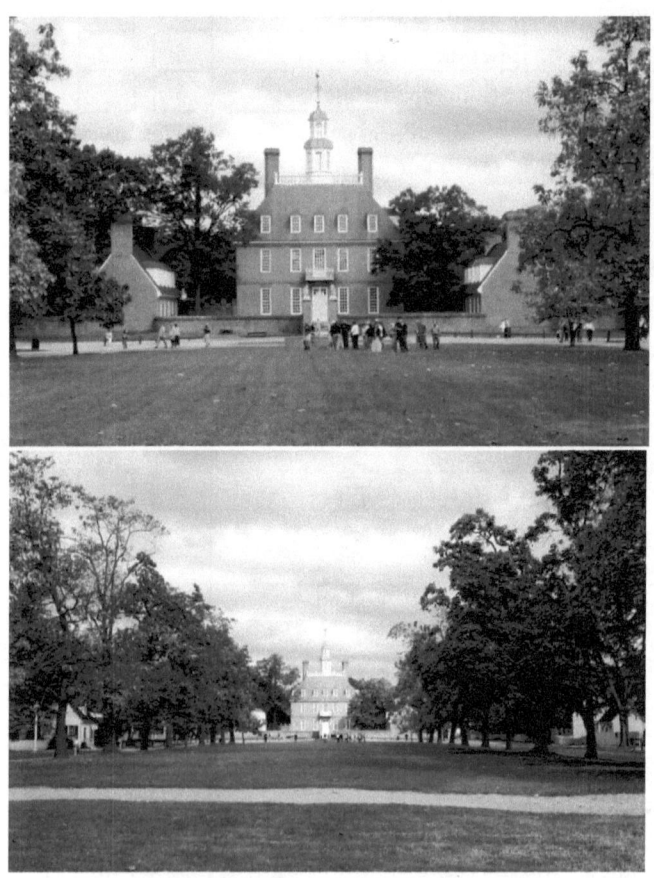

Views: **Looking North across Palace Green**

Recap of Five Voices Scores

Total score, History Density =	75	= 27% of Total Score
Total score, Intrinsic Symbols =	62	= 22% of Total Score
Total score, Cultured Symbols =	22	= 8% of Total Score
Total score, Bio-Kinetics =	63	= 23% of Total Score
Total score, Intention =	56	= 20% of Total Score
Total Score, All =	**278**	(of possible total 375)

Notice that two of the five clues posted the lowest scores, Cultured Symbols, and Intention. These surprised me, for this period of architecture is generally loaded with Cultured Symbols and high intentions. And, there were many Cultures Symbols in this restored city, but there were very few story reinforcements around the exterior of this particular example for us to grasp. A Cultured Symbol without assertive story (meaning) maintenance soon dies.

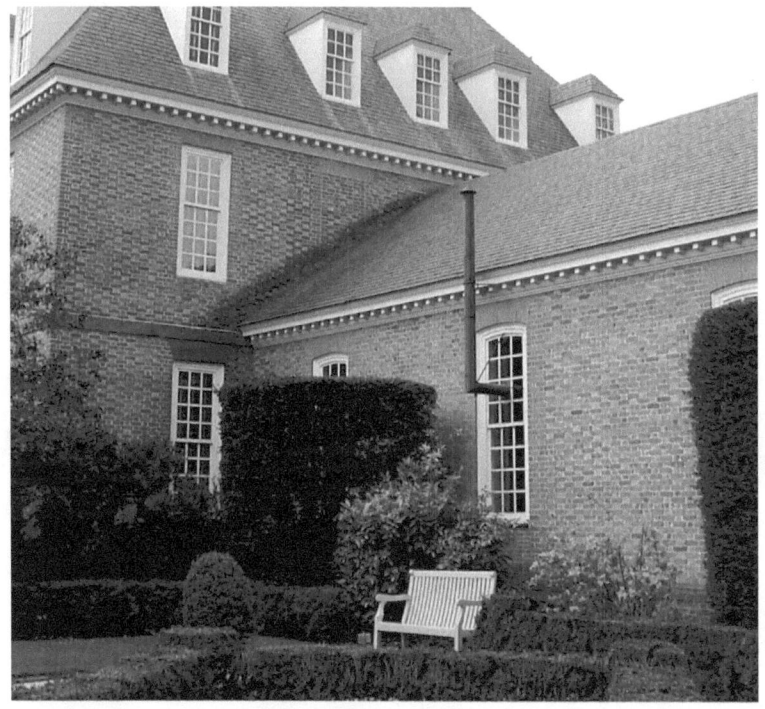

Reckless Intentions

The photo above illustrates one of the reasons that awareness training of all Five Clues should be made mandatory for all facility staff, including maintenance and engineering. Awareness helps facility staff to realize that each of them essentially carries special toolboxes with wrenches that speak the languages of each of the Five Clues. This photo also illustrates one of the reasons I scored the particular place example low in intentions. Do you see any other apparent intentional misfits this back addition causes?

7

CLUE FOUR - BIO-KINETICS

The process of human responsiveness triggered by the dance of the eye following the path of highest attractions – raw chorography for the eye.

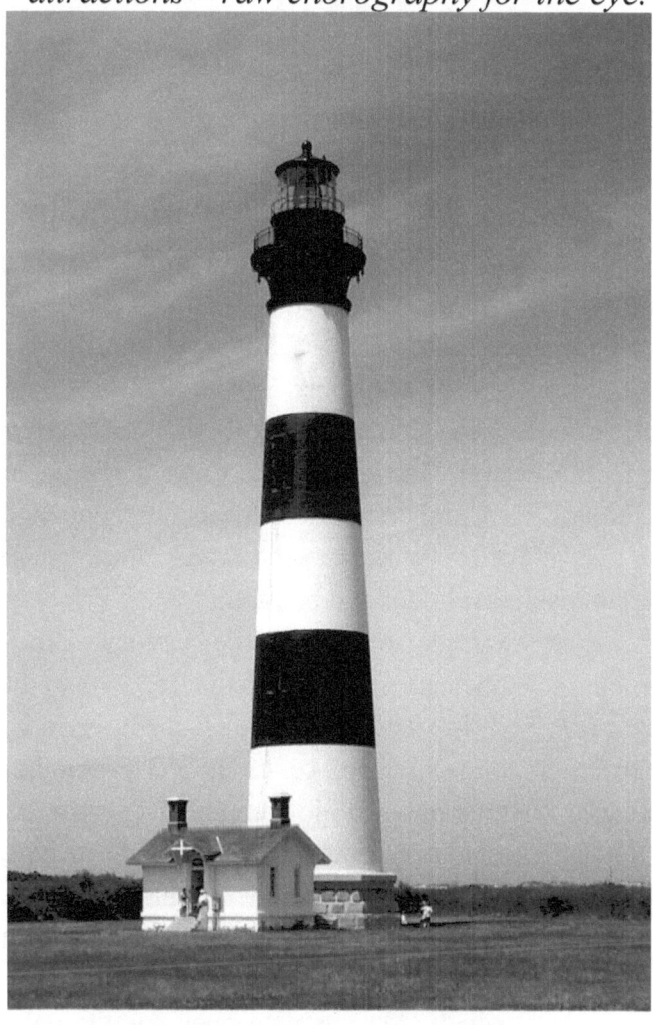

Bodie Island Lighthouse, Oregon Inlet
Near Nags Head, N.C.

The Bio-Kinetics of a circle is wholeness, the same response that is generated from the Intrinsic Symbol circle. Yet the Bio-Kinetics of a composition which contains objects aligned in a circle is honor – honor towards the point upon which the objects rotate. The objects do not have to complete the entire circle in order to generate the response of honor. Only enough of the arc is necessary to unquestionably demonstrate that alignment around a point or points has been achieved – and thus a sense of honor towards the focal point(s) is reinforced. A sense of bonding among the objects is also radiated through their acknowledgement by position of their shared alignment.

A popular Bio-Kinetic arrangement

A good example of this is found in a simple neighborhood cul-de-sac, with houses grouped around a circular street ending or a circular loop road. A sense of honor is generated by the grouping, while a sense of communal bonding between the houses is shared through their similar physical alignment around the common focus point(s). Intrinsic colors used by the houses will enhance the honor. Unrelated colors, forms, materials and other ingredients in the circular composition will dilute those Bio-Kinetic relationships of honor and bonding.

An ancient example of this same circular Bio-Kinetic phenomenon is Stonehenge, an approximately 100' diameter stone monument built in stages on the Salisbury Plain in England from the twentieth to the sixteenth centuries B.C. Thought to be a huge compass, calendar and cosmic computer of season patterns, Stonehenge and similar megaliths are also thought to have been used as sacred precincts for religious rituals. Stonehenge's circular composition is made up of 30 giant vertical stones set along the circumference and spaced apart from each other (slightly less than their widths) leaving openings through which the sun's and moon's positions would appear in alignment with the center of Stonehenge at various seasonal stages. Placed across the vertical stones are mammoth cap stones. One can imagine the suitability of this ancient technical place for religious ceremonies as well. Imagine standing inside this stone ring surrounded by the sense of honor demonstrated by the original 30 (now only 16 remain) upright stones orbiting the center. Couple that with its own history density and the cultured symbols of its stories of ancient cosmic scientific purpose, and one begins to imagine the awe and ennobling qualities generated. Visitors today still confirm these feelings. Many describe it as an ancient Thin Place.

This fourth process through which we respond to our visual environment is a human sensorial reaction process I term **"Bio-Kinetics."** This is the most difficult of the Five Clues to explain. One of the reasons is that its effects are the least perceptible consciously, particularly by the untrained eye, yet this response mode, this clue, is the most popular and widely used design tools of the five.

Popular yet difficult to understand

There are other paradoxical characteristics of Bio-Kinetics as well. While it is the most frequently used design tool by artists, designers and architects, its actual working methods are rarely discussed, quantified and even understood. Often, the bases of its influences are passed off as opinion or fluffy aesthetic justifications, yet its foundation is objective and essentially medical in nature. Access to the Bio-Kinetic library of tools, like those used in the initial stages of music composition, art and design, is usually through intuitive "flow" rather than through rational steps of reasoning. Yet the workings of this influencing process can be synthesized into working parts, like musical scales and notes, to foster study and evaluation to help understand its nature.

There are two key benefits in developing a basic understanding Bio-Kinetics: First, an understanding of the science of Bio-Kinetics can help those on the team of facility design, to develop and manage for better participation, cooperation and guidance in the creation, support and maintenance of the facility. And, with some discernment of the other four response clues, we can better differentiate between a poor design and a successful one – on paper (on screen) and ahead of constructing the facility.

What's behind its name?

As a term, Bio-Kinetics illustrates both the process and the complex visual digestive system supporting human responses through biological reactions (bio) and to movement (kinetics). Some derivation notes from the <u>American Heritage Dictionary</u> may be helpful. Bio, as meaning life, a living organism, (Greek bios, life) is used to signify influences upon our being. Kinetic (Greek kinetikos, as moving and kinein, to move) of, relating to, or produced by motion is used to signify the "engine" in this response process. Bio may also imply Bi as two, (defined also a variant of bio), bi as occurring twice, (Latin bis, twice). Under the category of perception, humans do in a sense, see twice, once as it appears, and secondly, as we are influenced (such as in an optical illusion). Bi also infers two, as two eyes which enables us to perceive depth. This is the general basis behind this response process: bi, bio and kinetics. Bio-Kinetics.

Bio-Kinetics is the process of connecting eye-movement and/or one's own body movement (kinetics) through space with internal responses (bio). It is separate from symbolic perception. It becomes or follows the motions of sight and converts the "ride" into a treasure of visual experiences and meaning. The continuous optical flight path of the observers eyes is kinetic, navigated by involuntary instinct, guided by form, light, color, texture and other elements which influence sight, from one visual receptive point to the next, from one column to the next, from the direction of one wall "looking at" another. And, on and on it goes until the design sequence concludes as planned, or falls apart.

This is the inherent melody of aesthetics. How we react to that adventure is the "bio connection." Bio-Kinetics is a very profound ingredient in the visual experience.

A simple Bio-Kinetic example

A simple example of bio-kinetics in action may be demonstrated through the selection and perception of fonts. Here is a simple example. Which is easier to read, the left or the right?

T   T

Many may select the left, because it is simple and clean and so it must be the easier one to read. There have been many objective studies done on this subject. Reading speeds and comprehension levels using different fonts have been measured. This is important to publishers and advertisers because advertisements and publications which are difficult to read reduce their income. Good fonts to publishers and advertisers are fonts which we can perceive easily and quickly. The answer to this question of which general font design succeeds here does not vary among educational levels or income groups or opinions. It is simply an objective fact.

If you guessed that the serif font on the right is easier to read, such as Times New Roman, etc., you are correct. Even though the other font is simpler in form, it is harder to read. If you wish verification of this, pick up any number of books on the art and science of font design, and you will see documentation and confirmation of this in the publishing world. Serif fonts are perceived more quickly and clearly than non serif fonts. Now here are two important questions for us. Why? And, what might this have to do with Bio-Kinetics and Thin Places?

Try this Bio-Kinetic test yourself

Close your eyes for a moment and determine (by feeling) the shape of your computer monitor on your desk. Go ahead and try it, then open your eyes. Remember when your eyes were closed how you rubbed your fingers across the edges several times, feeling its most extreme change in surfaces? Repeat the process to verify this. What you did was search with your fingers to determine changes and edges. Remember how much time you spent on those edges in comparison with the time along the surfaces? Understanding the edges is important in perception. We do the same with our eyes. We spend more time on the edges. A serif is a purposeful "over-design" of an edge to exaggerate that there is a change in direction, a change in surface plane. Serifs help us to spend less time on the edges, because it confirms quickly that there is a definite change here, and so we can proceed faster. And read faster. A serif is a Bio-Kinetic font edge tool.

Bio-Kinetic column tops

Relating this to architecture, consider the edges of forms, cornices, roofs, plazas, etc. Often the mistake is made in equating the use of serifs with symbolic representations of the capitals on columns. Yet, the reverse is true. Indeed capitals are Bio-Kinetic expressions of "how a column wants to end," just like they are on fonts, bell towers, fence posts and many other architectural elements. The visual "need" for such "help at the edges" is timeless and will never change unless our vision characteristics change substantially. The specific design within that Bio-Kinetic blob on the edges along the eye's travel path is where symbols often come into play. Remember from chapter three that we humans are addicted to symbolizing just about everything. So now here we go symbolizing the Bio-Kinetic need (and minor structural need) to top-off columns with capitals such as Greek and Roman column orders. The column orders are decorations of the edges. The design "needs them," and the reason that capitals are satisfactory is Bio-Kinetic.

We have been illustrating here so far Bio-Kinetics as triggered primarily by sight perception directing a shift in one's attention focus. And that sight route is a primary one indeed, but not for everyone, nor for all places. All of the senses can be a preceptor of Bio-Kinetics. As we sit in an auditorium, for example, listening to a speaker tell us about a recent trip, our eyes and ears keenly focused on the speaker. Then the air conditioning cuts on and the resulting gushing noise from vibrating vents 10 feet overhead now shifts our attention focus.

You can perceive Bio-Kinetic moving shifts in focus through all of the senses. Look, listen, feel, smell and even taste Bio-Kinetic influences in your own environments. Imagine sitting in the shade in a park – reading. Soon the sun's angle shifts position and moves light patterns across your bench. You begin to feel the warmth on your face and hands. Your eyes sense the brightness, another Bio-Kinetic shift in focus. Now while you are reading this, look out of a nearby window. If there is daylight outside, notice the trees' leaves gently moving in the breeze. Notice the sun's light patterns and the bird flying by. Have you ever listened to the tapping sounds that rain makes on forest leaves?

Are your perceptions essentially Bio-Kinetic or symbolic or both? Consider particularly unpleasant views too. Are they bio-kinetically uncomfortable? Do your senses' paths remain in concert along the way…to an appropriate resting place even for a brief moment? Or, instead, does the path fall apart, leaving your eyes or ears wandering? Notice particularly bothersome areas. Does your eye movement paint the volume of spaces, or does it skip from element to element, pausing at the edges, and then is guided to move on to the planned next point? What is happening to your eye movement there? Does it sing? Or is it more like a cell phone out of range? Then, visit a building or place you adore. What happens there?

Places without people would be different

What about machines...where the design and decisions are a direct expression of function? Aren't buildings machines for living? Yes, in a sense they are. Technology is an ever increasing part of facility design today. But, unless we remove people from the recipe, there is no logic or sustainable economic justification to substitute pure technology-based answers for our environmental designs. If we remove people from the designed element, design is substantially simpler.

For example, consider the fuel injection component in internal combustion engines. The fuel injector (or carburetor for less sophisticated engines) works with a relationship among four primary elements (three for a carburetor), fuel, air, the engine and electricity (for the fuel injector). The design of the fuel injector is based primarily on these relationships, although other relationships are present and influential, such as economics and time. Not included in these primary relationships in this example, however, are people. Other than people depending upon the fuel system to carry out its function, there is no direct relationship with people. There is little human interaction required or considered in its design or its operation. If however an active relationship with people were to be required and become a part of the design criteria, the design would evolve into a very different result. Part of its function would change. Thus, to recognize and satisfy those added functions, the design would have to reflect additional decisions.

Designing people-occupied facilities around only static components is a wasteful exercise and by definition is an incomplete system, the effective economic life of which is a direct function of how long the cultured symbolic trend can be perpetuated. The same component design expertise can be accomplished in a facility that also recognizes human responses and includes comprehensive design decisions to satisfy all of those components, including a major one, which is human. This is how the space shuttle was designed.

Thin Places are of people

Becoming aware of the Bio-Kinetic process is like learning the musical scale. It is helpful in describing and communicating music, but its understanding is not mandatory in the creative process. The same is true with Bio-Kinetics. Most artists and designers, like musicians and poets, work through a creative process which they themselves may not understand. They simply house it and nurture it. Pablo Picasso said, as translated by Brewster Ghiselin, "The painter passes through states of fullness and of emptying. That is the whole secret of art."

Mozart was said to have written, as translated by Edward Holmes, "When I am, as it were, completely myself, entirely alone, and of good cheer – say traveling in a carriage, or walking after a good meal, or during the night when I cannot sleep; it is on such occasions that my ideas flow best and most abundantly. Whence and how they come, I know not; nor can I force them…"

Walt Whitman said, "The greatest poet has less a marked style and is more the free channel of himself."

Artists of all mediums become channels of creative flow. When we react to those works of environmental art, architecture, landscape architecture, graphics, signage, and other visual stimuli, we do so through a different process, but it also is one we may not understand. Understanding the mechanics of the process is not necessary for reaction and response. Understanding the response process helps to evaluate more objectively how successfully the works achieve the desired levels of stimuli.

I selected the lighthouse as an example of Bio-Kinetics. Although the lighthouse actually scored higher in another clue, it is still a good expression of pure Bio-Kinetics. Here are the response subjects I found popular in my reactions to Bio-Kinetics.

Bio-Kinetics – measured response subjects

<u>Motion – foreground, background</u>: Lighthouses have prominent vertical masses and lines. They are seen up close and at great distances, usually rising above neighbors and against an open sky.

<u>Motion in alignment</u>: Alignment is the magic word for lighthouses. Their height is so extraordinary that it appears to be not only warning from way up there (its original purpose) but also "looking out." As noted in my poem which follows the matrix score sheets, "it paints as it searches." Alignment with something out there is very obvious and strong. I gave this line item a perfect score of 15.

<u>Motion between angles, curves</u>: As lighthouse structure and light chamber are round, it sets up its own motion around itself. This is not as strong as if there were multiple tall structures together, looking back at a center, as we will see in the next example.

<u>Motion as one moves around and through</u>: For those who seek lighthouses which still allow public ascensions, I recall climbing one as a "must do once" experience. There are three great views in the process, one looking up from the bottom-center, one looking down from the top-center, and of course the one looking out from the top.

<u>Motion in transparency</u>: Not much here, because the lighthouse is opaque, except at the light chamber.

I was a little surprised at how high it scored on the Bio-Kinetics matrix sheet. But I wasn't surprised at all at the score in the Intention Clue. Another perfect score of 75. That, I believe, is why people particularly adore lighthouses. The fifth Clue, Intention which is the next clue.

Place: **Bodie Island Lighthouse, Oregon Inlet near Nags Head, North Carolina**

1 - History Density

	A	x B	= Score
Sense of long history here	4	2	8
A connection with me I can't explain	4	3	12
Feelings of wanting to stay a while	3	2	6
Urgings to be quiet and not talk	3	2	6
Feelings of something special here	3	2	6

Total score, History Density = **38**

2 - Intrinsic Symbols

	A	x B	= Score
Space - identification	4	2	8
Light - presence/contrasted	4	2	8
Proportions - intrinsic	3	2	6
Shapes - intrinsic	3	2	6
Color - intrinsic	3	2	6

Total score, Intrinsic Symbols = **34**

3 - Cultured Symbols

	A	x B	= Score
Stories anchored to this design	2	1	2
Stories behind forms and shapes	3	2	6
Stories behind layout	3	2	6
Stories - accessories & furnishings	2	1	2
Stories - colors used	1	1	1

Total score, Cultured Symbols = **17**

4 - Bio-Kinetics

	A	x B	= Score
Motion-foreground, background	4	3	12
Motion in Alignment	5	3	15
Motion between angles, curves	5	2	10
Motion as one moves around-through	4	3	12
Motion in transparency	3	2	6

Total score, Bio-Kinetics = **55**

5 - Intention

	A	x B	= Score
Of its being & purpose	5	3	15
Expressed in shapes-form-placement	5	3	15
Expressed in color and honor	5	3	15
Expressed in Bio-Kinetic dialogs	5	3	15
Expressed in continuity	5	3	15

Total score, Intention = **75**

Thin Places and Five Clues - Matrix

Place: **Bodie Island Lighthouse, Oregon Inlet**
       near Nags Head, North Carolina

A = Presence of this element in the composition, 1-5 scale
B = Quality Value or Story Maintenance on a 0-3 scale

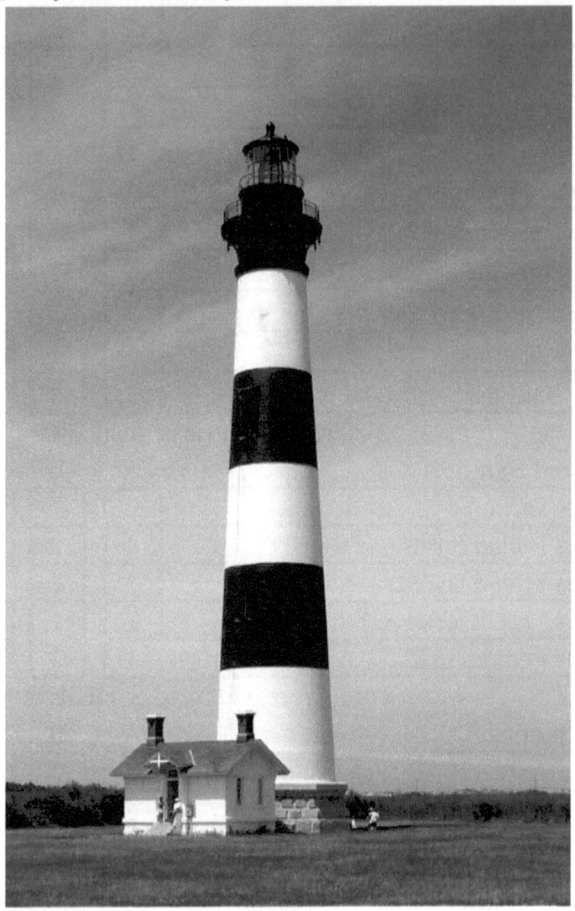

View: **Looking towards Eastern waters**

Recap of the Five Clues Scores

Total score, History Density =	38	= 17% of Total Score
Total score, Intrinsic Symbols =	34	= 16% of Total Score
Total score, Cultured Symbols =	17	= 8% of Total Score
Total score, Bio-Kinetics =	55	= 25% of Total Score
Total score, Intention =	75	= 34% of Total Score
Total Score, All =	**219**	(of possible total 375)

Walking Labyrinths

This Chapter on Bio-Kinetics at Thin Places would not be complete without a review of Walking Labyrinths which have roots in the practice of pilgrimage and their communicative potentials stored primarily in Cultured Symbols and Bio-Kinetics.

Every major religion practices pilgrimage although meanings and purposes vary widely. What is pilgrimage? Simply put it is a planned walk (usually long) to a very special place, maybe a Thin Place or a shrine or to a special person's place. The complete pilgrimage process contains three stages. The first is the trip there; the second is the actual experience while there; and the third is the trip home. In Celtic history many of those places visited were Thin Places. The pilgrimage process, however, does not necessarily involve spirituality, although historically most were such journeys.

A pilgrimage example today

A contemporary example is planning an automobile trip out of town to visit a beloved aging aunt who is in the hospital. During the trip to the hospital (stage 1) the driver might reminisce on how friendly and loving that aunt has been. Certain inner questions and preparations during the trip might come up. What is the best way to honor her in her current state and show her our love? What gift might 1 bring her? This driving there (stage 1) includes preparation for the actual visit with her. If spirituality is included, prayers may be offered in the aunt's behalf along the way. There may be stop-off points to reinforce the preparations, chapel visits and prayers for a safe trip and for the aunt to improve her health.

Then upon arrival, stage 2 begins with the actual visit which could last hours or days. After that, driving home (stage 3) may find the driver reflecting on what went on during the visit, wondering what to take back to others at home, and what changes will be evidenced in the visiting driver's life as a result of the visit. This is a simple form of pilgrimage.

In early Christian Celtic culture; imagine a family making an annual pilgrimage to a favorite Thin Place, worshipping there and then returning home. It is a long trip. Legend has it that walking labyrinths were used to symbolize (a Cultured Symbol in itself) that pilgrimage walk, and may have been used by aging family members who could not otherwise make the real trip. So the labyrinth was built at home or at a nearby special place and people walked its circuits to the center of the labyrinth, as a symbolic pilgrimage walk while others were on the road towards their distant destination.

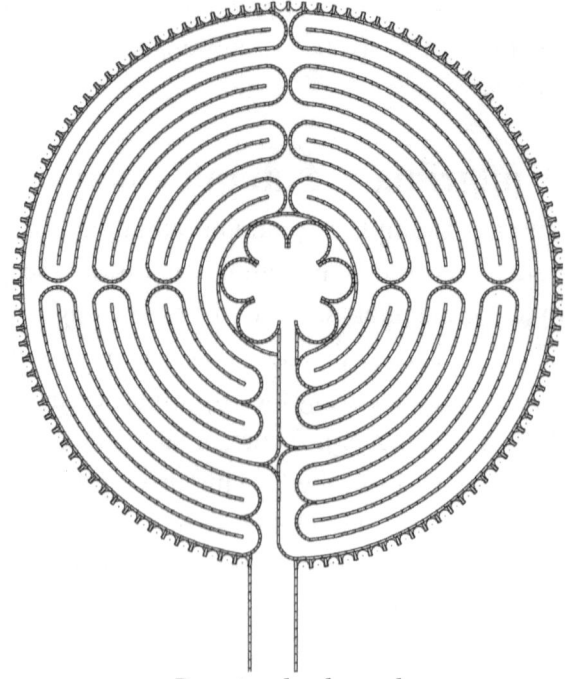

Drawing by the author
An eleven-circuit single pathway walking labyrinth similar to the layout of the floor labyrinth in the Chartres Cathedral in France. With your pen, start at the entrance and follow the path (stage 1) to the center, and pause there (stage 2). Then follow the path out (stage 3).

A key reason I believe those early labyrinths engaged the users was they were walked by and probably invented by people who also made the actual walk – that tough long tiresome and often dangerous walk. They had a powerful story or stories of the real walk to anchor to a hometown smaller version, a Cultured Symbol called a walking labyrinth. Keep that point in the back of your mind as we take a focused look at labyrinths today.

When those early Celtic labyrinth walkers used the circuitry of their particular labyrinth design, they must have remembered and re-experienced some of the same feelings they had in their younger years during land trips. I can imagine them also collecting and distributing memorabilia of the real journey, stones, vegetation from stop-off points which may have been transplanted along the labyrinth paths, maybe at the turning points. This is a beautiful application for us to consider for Cultured Symbols. Objects which are anchored with experiences can become triggers for recalling those experiences (stories).

Now, actually walking the pathways of a labyrinth is also a good example for us of Bio-Kinetics in action. For, as we move along the circular paths, it becomes bio-kinetically obvious to us that the circular arcs are aligned around a center. In this case, that is the center area of the labyrinth. That center area also represents our destination, the focus of our symbolic trip, the shrine or the Thin Place miles away. Some have called the circuitry of the Chartres Cathedral labyrinth pathways a spiral but that is incorrect. They are concentric circles. And even the walking process does not move in a spiral-like pattern to the center. If you trace your pen on the pattern drawing of the Chartres Cathedral, you will notice that the bio-kinetic movement does not move closer and closer to the center, as a spiral would. Instead, it moves in, then outward, then in, and so on.

To me the relationship of the non-spiral travel path in the Chartres Cathedral labyrinth to an actual pilgrimage trip presents a design stumbling block. A true spiral process, whereby the walker gets increasingly closer to the target would seem to be a stronger metaphor and a stronger Cultured Symbol of the real walk. A spiral process also would strengthen the Bio-Kinetic responses with the walker sensing his/her movement as journeying closer and closer to the aligned center. In addition, a spiral shape is also an intrinsic one.

A very notable book on the history of walking labyrinths is Walking a Sacred Path by the Reverend Dr. Lauren Artress, published by Riverhead Books, 1995. Dr. Artress is well known for her tenacious work in spreading knowledge and practice potentials with canvas labyrinths all across the world. I met Lauren in Richmond years ago when I was working on an outdoor labyrinth project. As Dr. Artress says in her book, "Labyrinths should be anywhere we might go in pain, confusion, reverence, or celebration." Truly, the labyrinth is a tool that can help us connect with an inner quiet. Sometimes it can be a Thin Place.

Why Labyrinths are – and aren't Thin Places

There are hundreds of labyrinth designs. The most popular in the USA is the one patterned after the floor labyrinth in the Chartres Cathedral in France, primarily due to Dr. Artress' fine and extensive work. But that is not the only design and some others may be more appropriate for your particular situation. Alex Champion, author and builder of mazes and labyrinths, wrote an important book entitled Earth Mazes, Earth Mazes Publishing, 923 Polk Street, Albany, CA 94706. He illustrates dozens of labyrinth designs. His book is also referenced in Dr. Artress' book. There are generally two schools of thought on labyrinths design. One is rooted in what is referred to as Sacred Geometry. The Chartres Cathedral labyrinth was apparently designed around that vernacular. The other school of thought, as evidenced in Champion's work, is a more sensorial approach, a "commodity, firmness and delight" approach, where designing elements, pathways, etc. are based on what one wishes for the user to experience at those points – similar to how most buildings and landscapes are designed today.

My preference perhaps borrows a little from each, but relies primarily on the Five Clues found among Thin Places. For now, thanks to Dr. Artress' re-awakening us to the engaging possibilities of multiple labyrinths in our communities, I believe it is within labyrinth sponsors' grasps to go for labyrinth Thin Places.

Surely, we must remember that the labyrinth is a Cultured Symbol of a very different actual road trip. If we are not strongly guided by new stories and associations with other three stage pilgrimages prior to our labyrinth walk, the walk experience may become meaningless. Simply recalling the history of the labyrinth itself will probably be insufficient to 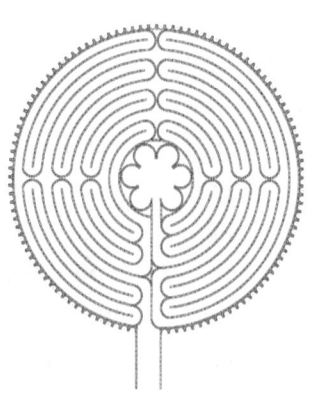 catapult you into relating to the three stages process with what is going on in your life now. And, if the Cultured Symbols do not have sufficient anchored stories, and their meanings have not been actively maintained, the symbol will die – and users will respond to the walking labyrinth intrinsically and Bio-Kinetically.

Remember harness horse racing?

I am reminded of the days watching harness horse racing on TV with the horses classified as either a trotter or a pacer, pulling its jockey in two-wheeled a light-weight buggy. The horses were trained to run according to the cultured race rules as if they had no joints in their legs. Every so often during a race, you would see one of the highly cultured horses forget and "break" their trotting or pacing in favor of their natural (intrinsic) gallop or cantor. The rules required them to then fall back and let others pass, or be immediately disqualified from the race.

This is similar to what happens to us when we respond to a Cultured Symbol (trotting) for a while, but soon forget the story (training) behind the symbol. We break into our natural (intrinsic) gallop. If that happens at a labyrinth walk, the labyrinth will be perceived intrinsically and the walker may lose focus and the experience of a pilgrimage. Other designs, of course, are possible as might be developed by your architects using what we can learn from the Five Clues within the architecture of Thin Places. If a Thin Place experience is the ultimate goal of a walking labyrinth, and I believe it is, then why not also consider all of the five Thin Place clues? There are more successful Thin Place examples in the world of <u>various</u> designs than there are of exact duplicates of the Taj Mahal.

If there is such a thing as "sacred geometry" most of the architects today don't use it extensively, perhaps because they have found or have changed the name of better and more meaningful design criteria. If there is a sacred geometry, I suggest it would be "Intrinsic Geometry," and found in nature as another example of the Cosmic Truth around us. Intrinsic Geometry in nature never dies because its connective influences are hardwired into each of us. We have already seen that the Golden Section ratios, spirals and circles are three intrinsic examples found in nature which are particularly appropriate for use in labyrinth design.

Conversely, I submit that any "sacred" geometry elements which are not found in nature are instead elements of a cultured geometry (cultured symbol) rooted in the minds of humankind somewhere in history or perhaps in someone's drafting room. Unfortunately, I have seen this infect some who consider building a labyrinth but insist on repeating – perhaps out of confusion or fear – certain exact contrived angles, pointed stars or other magic rules that cannot be supported by multiple ancient or recent Thin Places nor by professional designers nor by a consistent presence in other highly engaging places.

A labyrinth Five Clues score sheet

As an exercise and example, I used the Thin Place scoring matrix to score my experiences during a number of walks at three different labyrinths, two of them indoors at two local churches in Richmond, Virginia while the third one was outdoors. All three were of similar design as the one at Chartres Cathedral. Your experiences at these labyrinths may be substantially different than mine, of course.

Although I have been a fan of labyrinths for years, generally my actual labyrinth walking experience has been somewhat disappointing to me. Like many others, I can meditate or experience pilgrimage most anywhere. It does not necessarily have to be in a quiet place or at a Thin Place. They help, but they are not absolutely necessary. However, when I visit a place that is intended to be a tool to assist, enable and encourage some internal meditative processes, but it fails to engender little more than a walk in the park, my "fix it" nature and architectural training nudges me to investigate this a bit further. My labyrinth score sheets which follow summarize my perceptions.

History Density: Of course a portable labyrinth is not a place. It is a covering over a place. History Density is a response to the history and intensity of people previously at a place. If the portable labyrinth is in a church gymnasium, the history density may instead reflect last week's basketball game or last month's art show. I know the history of labyrinths, but knowing the history of a projector screen does not help me understand the movie.

My labyrinth experiences did not sense dense layers of people like I did as expressed in my poem "Through the gate' or at the plantation chapel, or even at the lighthouses. My experience at the outdoor labyrinth was not much better because the history of the site is substantially different than the labyrinth. Even so, I allowed the history of the labyrinth to be a part of this evaluation and I probably ranked it higher than I should have because I truly like and honor what the sponsoring folks are trying to do.

Intrinsic Symbols: Other than the partial circles, there are very few intrinsic symbols in the labyrinth pattern.

Cultured Symbols: The labyrinth itself, of course, IS a Cultured Symbol, but the story maintenance presented was primarily written history of the symbol rather than detailed stories on how people used it and reacted to it. Perhaps a play with three or four characters representing historical users of labyrinths during different periods of history would help labyrinth users get a deep feeling (story anchor) of the "real" walk. Maybe a video of such a play watched by users ahead of the walk would help.

Other anchors within the labyrinth may also help connect the story with the users such as relics or photos of relics or Celtic symbols and other devices used in the play, all parked along the labyrinth pathways.

Bio-Kinetics: This received the highest score from me, 64 out of a possible maximum of 75. Not bad. Lots of motion, and I may have even scored the Quality Value of the motion a little higher than I should, because actually there were very few "receptors" of the motions to carry meaning. For example, as I walked around the circles, I became aware that the arcs I was making had a single radius point, and it was there in the center of the labyrinth, and I knew that I would eventually be there. This is good pure Bio-Kinetics, but as noted earlier since the pathway takes users in close to the center very early in the walk, then back out, then back in, this took away from the Bio-Kinetic potential of "spiraling" gradually getting closer and closer to the center.

Another good Bio-Kinetic opportunity which was missed by the geometry getting in the way was at the turning points where the walker must make a 180 degree turn to head in the opposite direction. See those on the layout? What a great opportunity to stop for a moment, perhaps to sit down and reflect on "turning points" in one's life-journey. But, there's not room to stop, or sit down at the turning points. This is another missed Bio-Kinetic receptor point.

Intention: What intentions do I perceive here from the labyrinth? One, someone cared enough about us walkers to make this available for us. Thanks. How about from the labyrinth itself? A similar intention of care and good work comes through to me as a result of Dr. Lauren Artress' efforts, her book and her travels to re-introduce the rest of us to the labyrinth. Then, I sense an intention from somewhere out there that we need to help others carry this work forward.

So the intentions come from three levels, the sponsors, Dr. Artress' work, and finally from the actual labyrinth here which I scored the lowest. Comparing this to other spiritual places and Thin Places, its intention and "thinness" here did not come through clearly for me. Perhaps it is difficult to feel connected with spiritual intentions under fluorescent lights or out on a parking lot. For the labyrinth to be more effective than normal "thick" places, it needs to be a set-aside well intentioned and designed total place that embraces the labyrinth, with all five clues as described here, working to contribute to its success. Here are my scores:

Eleven Circuit Labyrinth, similar to the design laid in the floor of Chartres cathedral, France, in the twelfth century

1 - History Density	A	x B	= Score
Sense of long history here	3	2	6
A connection with me I can't explain	3	3	9
Feelings of wanting to stay a while	3	3	9
Urgings to be quiet and not talk	4	3	12
Feelings of something special here	3	3	9
Total score, History Density =			45

2 - Intrinsic Symbols	A	x B	= Score
Space - identification	2	2	4
Light - presence/contrasted	2	2	4
Proportions - intrinsic	3	3	9
Shapes - intrinsic	3	3	9
Color - intrinsic	2	2	4
Total score, Intrinsic Symbols =			30

3 - Cultured Symbols	A	x B	= Score
Stories anchored to this design	4	1	4
Stories behind forms and shapes	4	1	4
Stories behind layout	4	1	4
Stories - accessories & furnishings	3	1	3
Stories - colors used	1	0	0
Total score, Cultured Symbols =			15

4 - Bio-Kinetics	A	x B	= Score
Motion-foreground, background	5	3	15
Motion in Alignment	5	3	15
Motion between angles, curves	5	2	10
Motion as one moves around-through	5	3	15
Motion in transparency	3	3	9
Total score, Bio-Kinetics =			64

5 - Intention	A	x B	= Score
Of its being & purpose	4	2	8
Expressed in shapes-form-placement	4	2	8
Expressed in color and honor	4	2	8
Expressed in Bio-Kinetic dialogs	4	2	8
Expressed in continuity	4	2	8
Total score, Intention =			40

Recap of the Five Clues Scores			
Total score, History Density =	45	= 23%	of Total Score
Total score, Intrinsic Symbols =	30	= 15%	of Total Score
Total score, Cultured Symbols =	15	= 8%	of Total Score
Total score, Bio-Kinetics =	64	= 33%	of Total Score
Total score, Intention =	40	= 21%	of Total Score
Total Score, All =	194	(of possible total 375)	

While there are potential Thin Place possibilities and benefits in walking labyrinths, many of the current applications I have seen tend to limit those potentials by allowing rigid geometry to preempt the inclusion of the five clues as found in many Thin Places.

What might architects do with labyrinths instead? I certainly cannot speak for them, but I encourage you to talk with them about their ideas. I also encourage you not to begin a labyrinth project without an architect.

Architects, with their clients together can move this dream ahead, perhaps by adding sculpture along the way; or if at a church, maybe by including religious symbols along the way with the cross within the labyrinth's center. For outdoor models adding landscaping and designing the center space with surrounding trees that arch over the center to enhance meditation and accommodating 12 people sitting in a circle there. Consider also including pull-off places somewhere along the pathways and at the 180° "turning points." And most importantly, design the set-aside labyrinth place to radiate the intention of prayer and the mission statement of the church or the sponsoring organization.

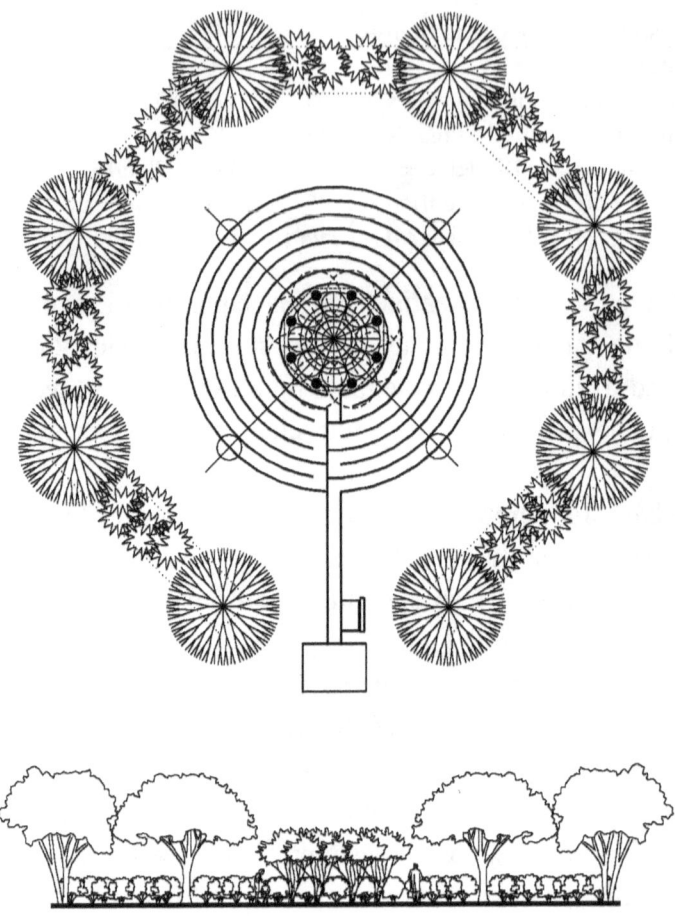

Conceptual sketches by the author illustrate some variations which can be made for labyrinths to suit specific objectives. This "set-aside in its own place" 7-circuit outdoor labyrinth contains a substantially larger center to accommodate 8 to 12 people. Importantly too walkers move in a spiral-like route as each circuit follows appropriately closer and closer to the center. Note also 4 pull-off stations along the first circuit for preparatory pauses and/or sculpture locations. In addition to the space-defining 8 trees and shrubs encircling the labyrinth, 8 select smaller trees (shown with dashed lines) arch over the center to enhance meditative focus. Many other changes are possible. Talk to your architect about designing a labyrinth to best suit your goals.

8

CLUE FIVE - INTENTION

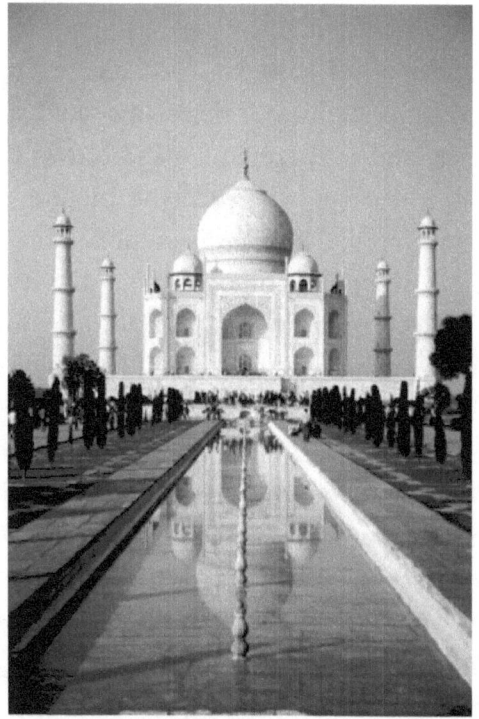

Taj Mahal, Agra, India (*Getty Images*)

There is a fifth and distinct clue to how highly engaging places and particularly Thin Places capture us. That is through our perception of Intention. Intention (from <u>The American Heritage Dictionary</u>): "a plan of action; design. An aim that guides action; object. The general connotations or concept of something; what something is meant to convey."

We perceive intention in conversations, don't we? As we do, it affects our response to what was said. If we perceive that the words uttered to us are intended to undermine our own efforts, we react quite differently than if we perceive the same words uttered with the intent to compliment and support. One of the difficulties in communicating with written words is that the readers cannot see the gestures, facial expressions and other signals of intent. In environmental design, however, those gestures and signals are abundantly revealed throughout the design, although the designers and programmers of the spaces may be unaware of the telegraphing flow of their intents.

Think about this for a moment. Imagine seeing a stream of trash scattered along a highway. There are bottles, broken jars, food boxes, paper, and garbage in a wind-blown row along the shoulder. Perhaps you can visualize the picture. Think for a moment about how you feel about this picture. Are you perturbed about it? What if it were trash dumped along the street in front of where you live? How would you feel about it? Hold onto those feelings for a moment and consider this. The trash itself is not the prime generator of any displeasure you might experience here. It is not the layout of the trash (bio-kinetics) nor even the symbols of each piece, as there may be beautiful pieces there as well. And, viewing the same trash in the dump probably would not produce the same irate negative responses if the trash is dumped in front of where you live. The main displeasure in seeing such a display along the road is the perception of <u>intentions</u> behind the scattering of trash. Intention is the primary response generator here.

In this example, the intention conveyed may be disrespect. Disregard for other people's property and observations. Intention perception, just as form perception, or story perception through cultured symbols is a part of our response system. Lack of honor is an intention easily perceived and seriously judged. Examples of vandalism and user disrespect researched in my practice have shown that "prison-type" protective finishes generate feelings of mistrust. Usually those glossy hard reflective "easy-to-clean" finishes get attacked first.

There are many negative examples and reactions to empty, contrived or incompatible intent standing in the built environment. Consider past facility user discontent in examples in which you are familiar. The intentions behind the building itself, may be a major part of the problem. Then the design which fulfills those facility intentions may be the remaining part of the problem.

Intension perceptions gone astray

One example is the architecture building at a major northeastern University, an award-winning outstanding design by then well accepted design standards. However, soon after it was opened years ago, the students purposely set the building on fire. Administrators and architectural faculty were shocked. Later, after interviewing the students, it was speculated that the design intentions which telegraphed to the students apparently were not compatible with their own need to freely experience and grow. The building design to them did not recognize or honor the students' wishes to freely express. To them the building overpowered their creative dreaming and didn't leave them room to naturally grow and develop. In short, this intention failure represented a miscarriage of the primary purpose the building was to serve. Intention is a very powerful prerequisite voice. Just as in verbal communications, if inappropriate intentions are perceived, the remaining words are often not heard.

Intentions should be programmed into design

What we should learn from that experience is that intention should be seriously reviewed and studied by the entire design team during the early programming phases of facility planning, well before any sketches are actually started. Intentions of the facility and spaces should be spelled out on paper, just like room size requirements. Are the facility's intentions a direct outgrowth of the users' intentions? Never assume they are known, because usually they aren't. Then, during the initial design presentations, a serious portion of the review and discussion process should focus on how the programmed intentions are being carried out in the design. Equally important in that review is to consider the questions: Does the facility design not only accommodate the program, but does it actually radiate and stimulate the program's deep intentions?

Perhaps the world's best example of the power of intentions

A magnificent example of the deep intentions of a facility owner being captured in a design, and then strongly radiating those feelings to visitors is the Taj Mahal, near Agra, India. One visitor wrote the following:

> "In a world of tyranny and cruelty, a heavenly dream crystallized in stone: the Taj Mahal. I cannot conceal my unmitigated admiration for this supreme flower, for this jewel beyond price, and I marvel at that love which discovered the genius of Shah Jehan and used it as an instrument of self-realization. This is the one place in the world where the – alas – all too invisible and all too jealously guarded beauty of the Islamic Eros has been revealed by a well-nigh divine miracle…It is Eros in its purest form; there is nothing mysterious, nothing symbolic about it. It is the sublime expression of human love for a human being."

The visitor quoted above was Dr. Carl Jung. (From C.G. Jung, World and Image, as previously referenced. Here the intention of love and honor in this memorial tomb for Shah Jehan's wife is so powerfully conveyed through the architecture, that it overwhelms its visitors, often bringing them to tears. It is interesting to note that C.G. Jung who might easily be described as the master of the language of symbols, said, "...there is nothing mysterious, nothing symbolic about it. It is the sublime expression of human love for a human being." And, while there are abundant Islamic cultured symbols used throughout the design, they are neutralized to Jung (and most of us) through non-Islamic perception. The resulting responses are generated through intrinsic symbols, bio-kinetics and the power of intention, artfully woven into form and space by the anonymous Islamic architect. Visitor responses of awe and soul-shaking admiration confirm this Thin Place each year, reaching across cultures and time throughout its almost 400 year history. I use this architectural Thin Place as an example in the scoring matrix. Here are the five responses subjects I found most appropriate for the Taj Mahal under this fifth clue, intention:

Intention – measured response subjects

Of its being and purpose: Overwhelmingly, the purpose of this place is communicated very strongly. That is to honor a beloved wife. Its architecture expresses that so intentionally and spiritually that visitors are overcome with emotion. Perfect score

Expressed in shape-form-placement: Perfect score

Expressed in color and honor: Perfect score

Expressed in Bio-Kinetic dialogs: See the four towers standing guard around the center. It is as if they are stepping back and saluting the center, the wife. They are bio-kinetically motioning that, and our eyes follow that pathway. Just as the dome rises up, and looks back down at her. Just as the reflecting pond invites us to step back and look at her, reflected here. Pure intentions of love and honor.

Expressed in continuity: Yes, another perfect score.

Place: **Taj Mahal, Agra, India**

1 - History Density

	A	x B	= Score
Sense of long history here	5	3	15
A connection with me I can't explain	5	3	15
Feelings of wanting to stay a while	5	3	15
Urgings to be quiet and not talk	5	3	15
Feelings of something special here	5	3	15

Total score, History Density = **75**

2 - Intrinsic Symbols

	A	x B	= Score
Space - identification	5	3	15
Light - presence/contrasted	4	3	12
Proportions - intrinsic	4	3	12
Shapes - intrinsic	5	3	15
Color - intrinsic	5	3	15

Total score, Intrinsic Symbols = **69**

3 - Cultured Symbols

	A	x B	= Score
Stories anchored to this design	4	1	4
Stories behind forms and shapes	4	1	4
Stories behind layout	2	1	2
Stories - accessories & furnishings	2	1	2
Stories - colors used	2	1	2

Total score, Cultured Symbols = **14**

4 - Bio-Kinetics

	A	x B	= Score
Motion-foreground, background	5	3	15
Motion in Alignment	5	3	15
Motion between angles, curves	5	3	15
Motion as one moves around-through	5	3	15
Motion in transparency	5	2	10

Total score, Bio-Kinetics = **70**

5 - Intention

	A	x B	= Score
Of its being & Purpose	5	3	15
Expressed in shape-form-placement	5	3	15
Expressed in color and honor	5	3	15
Expressed in Bio-Kinetic dialogs	5	3	15
Expressed in continuity	5	3	15

Total score, Intention = **75**

Thin Places and Five Clues - Matrix
Place: **Taj Mahal, Agra, India**

A = Presence of this element in the composition, 1-5 scale
B = Quality Value or Story Maintenance on a 0-3 scale

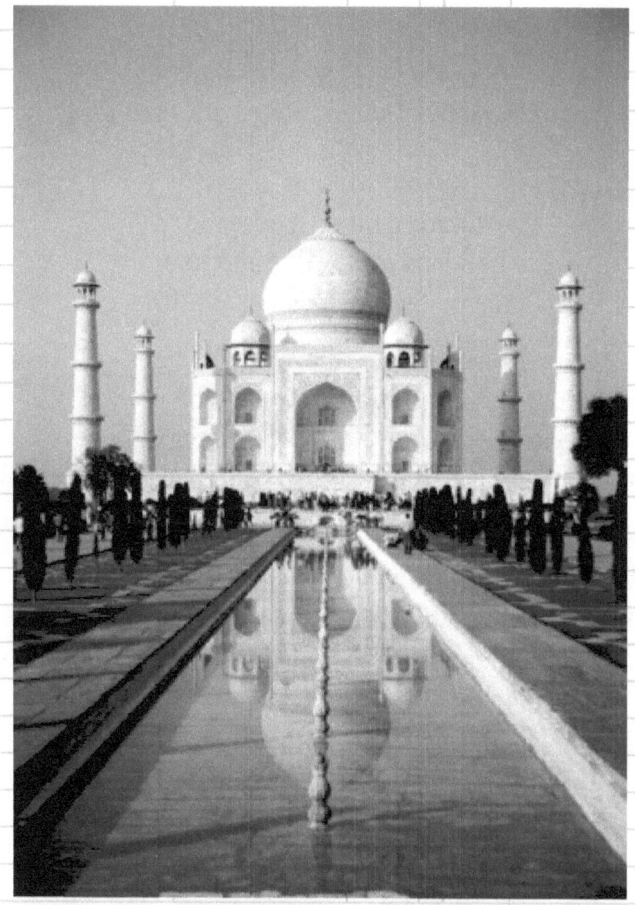

Taj Maha (*Getty Images*)

Recap of Five Clues Scores

Total score, History Density =	75	=	25%	of Total Score
Total score, Intrinsic Symbols =	69	=	23%	of Total Score
Total score, Cultured Symbols =	14	=	5%	of Total Score
Total score, Bio-Kinetics =	70	=	23%	of Total Score
Total score, Intention =	75	=	25%	of Total Score
Total Score, All =	**303**	(of possible total 375)		

This masterful beautiful design was no accident. Imagine Shah Jehan communicating to his architect how important it is to have this building radiate his love for his late wife? I conclude this chapter on Intentions with these two poems.

Winds at the Sea

An example through poetry of Intention

Like a Mother's love
floating on unseen wind,
Pelicans escort divinity here
across fauna's Disney World.

Invisible to us, the wind engages
in effects through what we do see,
as it paints on borrowed canvasses,
and thrives on routine unconcern,

to transform -- moments to magic,
walking to whisking waltzes,
while amusing itself in solitaire,
spinning new patterns in the sand.
For without wind, waves would
sleep humdrum, flat with the kites.
Balloons would sit out the dance,
grayed under clouds parked broken.

But instead, Sea Oats celebrate
standing ovations from wind's sway,
while sails beyond, puff deep,
flaunting breeze's catch of the day.

Still, phantom wind prays lonely.
Its palettes and engines linger,
hovering within its studios of whirl,
poised in yet another disguise, to
air-brush again our winged souls.

Concrete

A metaphor of people and intentions

Its recipe has three key ingredients,
water, cement and aggregate, each
carefully measured in proportions,
blended, then stirred and again until
a powerful chemical reaction occurs.

Water mixes with cement, saucing
a magical time-release glue, but
without lumpy aggregate, the glue
alone would cling vacantly to itself,
and bind potential out of control
like an engine without burden,
a person without mission, or
open wings without air, drifting,
coasting, wasting possibility.

But add lumps, as aggregate, graded
from the tiny particles of sand
to larger gravel and chunks of rock,
each swathed in glue-paste, stirred
until the potent catalytic trio is born,
a powerful assemblage, like people
bonded by purpose, love and unity,
and both mixtures are ready to build
solid foundations, beams, bridges
and lofty spans – yet unknown.

9

CONCLUSIONS
AND OPPORTUNITIES
Thin Places and Five Clues

We have walked through each of the Five Clues which churn human energies at Thin Places and other highly engaging places. We have reviewed how the <u>History Density</u> of a place can seize a person's awareness, at places like roadside flower clusters, the Rock of Cashel, places like the Virginia plantation chapel and others, including a number of places you may have experienced personally.

Then we reviewed <u>Intrinsic Symbols</u>, those symbols rooted in the Cosmos of nature and which never lose their connective power with us because we are hardwired to respond to them. They include light, space, color and shapes like the circle, spiral and Golden Section paired dimensions in the ratio of 1 to .618 found in abundance in the structures of humans, animals, plants, ancient art and architecture.

Next we reviewed <u>Cultured Symbols</u>, those symbols which are anchored to human story-experiences and which can produce powerful human responses – as long as the story itself is highly engaging and its presence in the memory of users is aggressively maintained. If the story slips from the memory of users, the symbol loses its meaning and dies. It is then processed through other means usually at a much lower or even contradictory level.

Then we reviewed the <u>Bio-Kinetic</u> Clue, movement affecting experience and perhaps the most popular architectural design tool which is driven by the skills and design talents of architects chasing after compositions to radiate human connections and intentions. Two very good examples of this clue-tool are found at the Taj Mahal and Thorncrown Chapel. Effective Bio-Kinetics cannot be adequately designed by mathematicians or historians, or anyone outside of an experienced architectural background and practice. Bio-Kinetics of place when powered by an experienced architect, best reach their potential.

Finally, we reviewed the fifth clue, Intention, probably the least used design tool today, yet in my experience one of the most consistent clues found at Thin Places. And, this may be a principle reason Thin Places are set apart and above "thick" places. Noble intentions of a place appear to be taken very seriously at highly engaging places and particularly at Thin Places. Consider, for example, the various ways that the deceased wives of leaders have been memorialized throughout history. I am sure all were properly acknowledged. Then consider how intensely the Islamic Shah must have communicated his deep intentions to his architect to show through a commissioned tomb design his intense love and honor for his deceased wife. The Taj Mahal does exactly that and visitors very strongly perceive that intention.

Then in a time and geography warp, in the woods of the Ozark Mountains, a retired school teacher built a simple retirement cabin. Soon friends dropped by to enjoy the scenery, and the school teacher had an idea…deep from within. Why not build a simple glass and wood chapel up on the hillside for his friends and all others to visit and enjoy the natural inspiration? He wanted to give back with what he felt inside. He didn't have a church to use it or pay for it, but he decided he wanted to do this for anyone and everyone anyway. That was his intention. (The full story is found back in chapter 1). Can you imagine him communicating that intention to his architect, the highly skilled Faye Jones? The rest is history. This leads us to the next concluding point.

Seeds for Thin Places?

As we look back on the examples of Thin Places used here and beyond – the Rock of Cashel in Ireland, Thorncrown Chapel in Arkansas, lighthouses of the world, the Governor's Palace in Virginia, and the Taj Mahal in India, do you see a common thread weaving its way through these diverse examples and others you may know? Think about them for a moment.

Here's the commonality that I see, which may be a beginning seed of Thin Places. Each of the examples appears to me to have at its core a seed of intensified focus (by its sponsor/developer) on honor and love. Look at them. Some of them have layers of seeds over the years. From the Rock of Cashel for more than 1,700 years and some for less than 50 years. Some are more history dense than others, but all have at the core intentions, of love, honor, protection and, for some, duty. None of them radiate an intention of opulence or lavishness, do they? Do we get that? None of them radiate an intention of "making money" though no one would complain over the economics of the Thorncrown Chapel.

More importantly, none of them radiate only an intention of just providing shelter or just providing housing, or just providing economical office space. They all have higher intentions. These, I believe, are the seeds of Thin Places and highly engaging architecture. So where does that leave us here, near the end of this book? I think it strongly suggests opportunity for all.

Opportunities to Seed High Engagement

Opportunity 1 – For all readers, please visit some of your favorite places and score your responses to them on copies of the score sheets provided at the back of the book here. Note, some of the blank score sheets have my response questions under each clue if you wish to use those. If not, some of the other score sheets have those blanked out as well, so you can add your own.

Scoring places will help you learn how to more objectively evaluate a projects influence and value – a very important, yet hardly ever performed review. You may be very surprised to see the patterns in your scores and actually see why some of them are your favorites, and why others appear to fail. Why build something that doesn't include the substance of your favorites? And, certainly why then build others that include the substance of human connection failures?

Opportunity 2 – For all of us who may be involved in future facilities development, including facility leaders, owners, officers or board members of corporate or institutional facilities, this study of Thin Places suggests that from the very beginning of planning a new project, we should work to uncover and develop the highest intention, the most noble purpose that this new or renovated facility should radiate from its very core. This will take some effort, but it is a huge opportunity to program deep into the facility the seeds which later will be interpreted and perceived by its users. If the highest intention we can muster is to provide more office space, then, that's what the users will read. I certainly don't see any special human spirited connection to that, do you? Instead, go back to the mission statement of the institution or sponsoring organization. Most publically accessible mission statements I have seen of organizations and institutions appear to have been well developed and thought out. Check those mission statements as a good starting point.

Once the detailed intentions are worked out, discussed and approved, they need to be communicated in writing along with the rest of the program to the architectural design team. Walt Disney did that so well. So did the Shah. Design progress presentations should review the proposed architectural solutions to achieve the desired radiation of these intentions.

Opportunity 3 – Learn how "conventional" procedures for facility programming and design often unknowingly exclude considerations for History Density and Intentions, two of the important five clues described here. Excluding these may prevent facilities from achieving the human connectivity level of high engaging facilities or Thin Places.

Opportunity 4 – If you are a spiritual leader, a minister or a lay leader of any religious group, please consider how your input and guidance on a next facility project is needed to help ensure that at least those two powerful spiritual influences are automatically built into the planning process – history density and intentions.

Opportunity 5 – If you are a "numbers" person and enjoy (like many of us) seeing the bottom line of otherwise successful ventures, learn how both the success and the bottom line may be significantly heightened by insisting that your next facility project include a goal of being a highly engaging facility, perhaps even a Thin Place. Remember, highly engaging places don't require renovations or makeovers every three or so years to address the next trendy whim, or to try again to correct empty intentions.

Opportunity 6 – If you are part of the facility maintenance staff, learn how you too are a key person in not only keeping the facility working, but also a key person in keeping the architecture and these five clues working. It's true. With you on the team and a supporter of highly engaging places, you can help keep the Intentions flowing, as well as keep the air flowing and the power flowing and the water flowing. Remember from the picture in chapter 6 showing what someone without engaging place knowledge which you now have, did with an exhaust flue poking through the (eyeball of a) window. What does that intention say to you?

These are opportunities for facilities to go really "green," to be fully sustainable, including sustainability of human spirit connectivity from the depths of their cores. Finally, I hope you may not only visit and score places as suggested in this model of five Thin Place clues – but that these five will also be a guide for you to bring that care and thinness into Your Own Places.

I transition these concluding thoughts through a poem about opportunity as I witnessed in one of our mutual feathered friends. Then, finally in the Appendix, I close with several other poems and essays.

Waiting for the Signal
To Seize an Opportunity

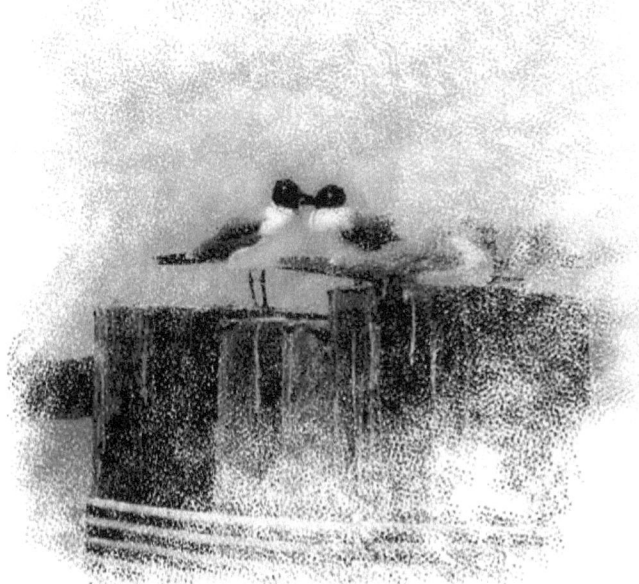

Hold still, proud one.
Audition for this poem, while
poised in your creosote office,
which harbors sleepy ferry boats,
and journeying shore commuters,
all stationed like you, perched.
All waiting for the signal.

Look me in the eye proud one,
connect our passing souls,
as if to find myself there,
reflected in your somber nod,
among your ignoring flock,
thoughts dancing in the wind,
while we wait for the signal.

Chirp soft your chatter,
while you tipi-toe ballet,
disguising positions,
and wiggles to be next in line,
like first-graders at lunch,
readying instincts,
waiting for the signal.

Ferry bells then chime.
Anticipation explodes into delight,
massive engines awaken,
floors shake and metal rattles,
children and grandparents giggle,
as we slowly creep forward,
launching off to journey.

A signal of opportunity for you,
proud one, as you squawk and dive,
flirting with killer propellers,
unfolding dazed fish in the wake,
extracted quickly and precisely
with swooping lunges, gulping beaks,
and dangling tethered feet.

Your predator rituals continue,
'til we dock on mirrored shore,
where I engage the spirit in your eye,
reflected in a miniscule wink, which
beckons from that placid pose,
to watch you again, linger
and listen calmly -- for the signal

APPENDIX

Metaphors of Thin Places and Spirituality
Expressed in poetry and essays

Friends

Turmoil pulls
On heart's strings
Laced in golden
Twine to Heaven,

While tangled
Strained threads
Often fall away,
Bruised and adrift,

'Til the yarn
Of friends
And the love
They weave,
Reconnect.

Seed Pods

They float, drift and fly
Through the crowded soup
Of suburban air,
Riding on invisible currents,
Some up, some down,
Some climbing to lofts unknown,
With gravity's fate lingering.

Tiny specs of life,
held aloft by dozens
of out-reaching rods,
each thin as a fly's whisker,
light as a spider's web,
Dancing in the wind,
With gravity's fate lingering.

One lands in the garden
Of such beauty to join;
It marvels alone in joy
While it proceeds to die
In dry shadows of the lush.

Another pod then descends
Floating like a parachute,
Pulling tight its chords to lift,
But falls instead upon barren land,
Far from lavish neighborhoods
And vacant of patrician splendor
where sprouts see only weeds,
but now these obvious needs,
is where magic soon seeds.

Rain Language

Harry didn't hear it,
just beyond his pickup,
at storm's finale,
engine shut down
windows cranked, waiting,

for George's truck
rounding call's bend
to rescue Harry again,
who ran out of gas
on this desolate road
eluding the squall.

Harry didn't hear it,
the rain splatters,
thumping thousands
of leaves, tapping
percussive passages
in flirting flora code.

As if talking through
dewing demi-drops
on levitating leaves,
wobbling wink-blinks
hurled at Harry, still
focused afar.

Then beyond clearing breezes,
which twirl leaves in boogie,
swishing swan's salsa,
signaling Harry, while seven
sparrows' a-cappella applause,
cinch again the cosmic cipher.
But Harry didn't hear that either.

Too busy to hear, to notice,
while chasing chosen chasms.
Until the next rain – when it
ambushed me again during a
fitness saunter along forest's brow.

Leaf Language

In May they suspend in silence,
Hushed green wiggles in the wind,
Which drone muted splatters
Under Spring rains, like
Glossy wet umbrellas.

Then October, drier they lull – tap, tap, tap
Against each other – graduated now
To hanging yellows, oranges and reds,
Which will parachute soon, searching
For a new place to hum.

But, it is late dreary November. When
Winded across cold-hardened streets,
Rattling dried brittle hollows,
Strumming those pebble-chorded roads,
They sing the loudest of all,
Celebrating still – the fall.

Wellsprings

You and I have been offered a unique observation point. From it we can see groups of people gathered around dozens of hand-dug wells, peering down the open excavations, straining to get a glimpse of the Great Overflow Dynamic, as it is called. It is known as the river of life. Deep below -- running through all of the wells and connecting each invisibly with a dynamic energy, the juncture of which is usually missed by the excavators above.

Each of the well sites has a sign designating its claim to the Dynamic. Some hold the sign high, proudly exclaiming the superior location of its site and thus the best access. Others use their sign as a shovel to scrape away the loose soil from around the opening. Others use the sign only to confirm their location.

We observe fights breaking out at some of the sites, people pushing and shoving each other, throwing stones towards other well sites which are believed inferior and a threat to the source Dynamic. Others strain to get into a better position near the deep void, to peer below – often loosening the soil and causing clumps and rocks to fall into the well -- splashing into the pure water below, and enraging the crowds into accusations and disciplines.

But, the pollution, as they shout at each other, is not realized to extend to all of the wells since they are all served by the same great source. Those at adjoining wells look over in disdain while lulling in comfort about the purity of their well. While the splashes of debris are not recognized as a temporary and natural process the river uses in its own dynamics.

There is an endless array of sites stretching out beyond us, some with large groups, some with just a few huddled together peering down into the shaft. Some with one person. Some sites are vacant...perhaps abandoned. And, some which apparently are only drawn to the crowds are clustered around dirt only. No well. Just a staked claim, in dirt for they see only holes at the other sites and not the source below. They yet see only dirt and empty holes. Yet the Great Overflow Dynamic energy still mingles below.

Murmurings abound, mostly honoring access points to the **Great Overflow Dynamic**. It has become symbolized among those gathered. Some refer to it as G.O.D. Others as simply GOD. Others portray it as the Great Spirit. The site signs are raised high in celebration as the groups cheer around some internal signal. One sign reads Jewish. Another Presbyterian. Independent Thinker. Buddhist. Islam. And on and on.

Yet, we can sense from our observation point that each expression is veritably sacred at its core. Not because of the fashion of their apertures in the earth. Nor because of their signs nor their proclamations, which often are against each other. Their loving sacredness emanates from each, in their own way, craving to tap the fringes, the hemline, the small bits of seemingly imperceptible and imponderable vapors of the Great Dynamic through their own crude wells. Our own wells. Pinholes in the vast infinite cosmic sea of GOD.

Your Computer Chips

I selected this essay as the final words in this book because it offers for your consideration questions that I pondered (still do) after each of my Thin Place experiences. You may have similar questions. I wondered afterwards in words something like this: "What is it about this place, this experience that is specifically aimed at me? Why? How can I best learn and invest whatever that is – to serve and be served through my own personal mission?

Imagine this. You are asleep and dreaming. In your dream, you have been selected to participate in a "Survivor" type game show. You find yourself alone on a small island with a heavy backpack and a single piece of paper with your next instructions. You read that each morning a small airplane will fly low overhead and drop three delicate light cardboard boxes to you containing a day's supplies and your next instructions. You are to build a device to catch the boxes without damaging them. The device itself as well as the salvaged condition of the boxes, and their potential use for others, will be determining factors in your success in the game. You are to mark your selected drop spot with the red blanket found inside your backpack.

Aside from the blanket, just about all else in the backpack is a knife and white rope – a big role of thin but strong rope. You sit down under an oak tree and begin wondering how you are going to do this. For some reason, you recall your high school geometry class. You remember your teacher explaining that a circle contains the greatest area within the least amount of material (perimeter). You wished now that you had paid more attention in class. But, you wonder if maybe your device should be a circle. Yet, how do you make a circle with rope? How do you fill in the circle so that the boxes won't fall through? Then, how can you support a circle above the ground so that when the boxes land, they will be cushioned by your device and not hit the ground?

Sitting there, you stare off in the distance and notice something glistening nearby between a bush and several pieces of scrap lumber piled next to it. You stand up and walk over to get a closer look, and you find a beautiful perfectly made spider web. It is about 8" in diameter. Right in the center is the builder, a tiny spider about the size of small green pea. Its head and brain are even smaller. Yet, here is this most magnificently engineered and efficient structure you have seen in some time – perfectly preserved for your observation.

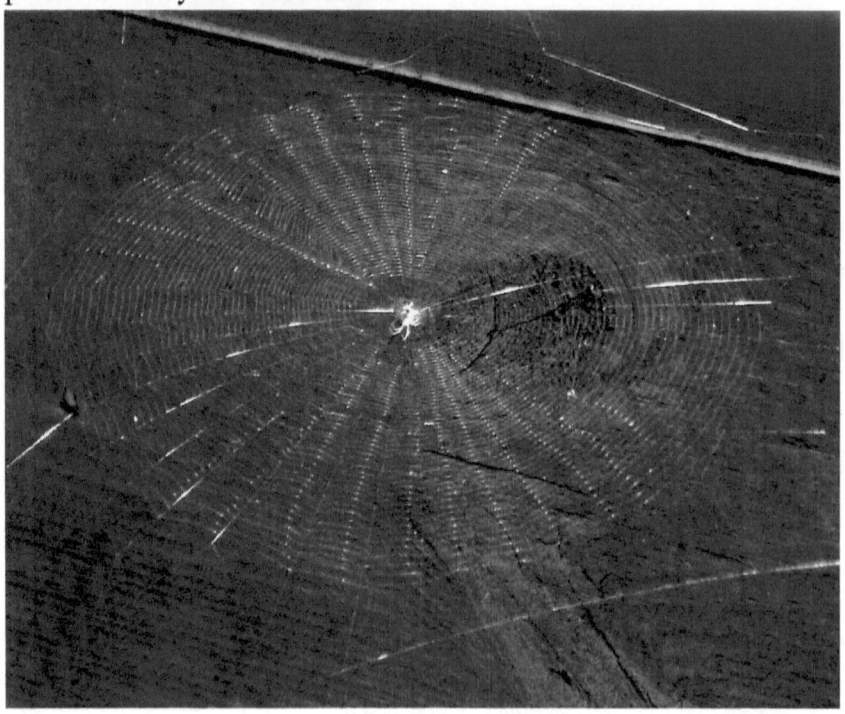

"…this most magnificently engineered and efficient structure…"

This may be an ideal model for your box-catcher device, you muse. You notice how the circular web is held aloft between two branches of the bush and one of the boards. There are runner "ropes" tied between the bush and the board at several points, then shorter ropes connecting the circle to the runners. Quite ingenious. So much so that your attention temporarily shifted to – how is this possible, with such a minuscule spider brain?

The spider is probably only weeks old. Yet, it designed and built this structural masterpiece – spanning the greatest area with the minimum material. How is that possible, when at 15 years old in geometry class, you could barely understand the circle? Something strange is going on here. Next, you hear music. Music?

Your clock radio awakens you back to reality. You stare at the ceiling for a moment, still wondering about that tiny spider and its ability to design and build such magnificent webs. As you peek across the room at another clock near the TV, you see that the time on that clock is different – an hour later than your clock radio. Then, you realize that you forgot last night to set the clocks to daylight saving time. The clock radio clock is correct, and the music-alarm went off at just the right time, because now you recall that the clock radio has a daylight saving time feature inside -- a computer chip which calculates the arrival of such switchover dates, even during leap years, and resets the clock automatically. Amazing. It was programmed to do that at the factory. You remember now, that all you needed to do when you first set the clock radio, was to push a "yes" button that gave you the option to activate the daylight savings time feature.

Your thoughts shift back to the spider and its beautiful web you dreamed about. Maybe that tiny spider is like your clock radio. It must have within its systems a spider computer chip programmed at the cosmic factory and implanted into the spider for survival purposes. Surely the spider's brain is not capable of designing and building such beautiful geometric and highly efficient structures by itself. But with an implanted web-chip from the cosmos the spider's brain simply goes along for the ride and watches for predators.

Are we humans like that too? Many think so. Do each of us have at least one of those special-purpose computer chips implanted in us? As we search for meaning in our lives, one of the things that we seem to learn sooner or later is that we too were "programmed at the factory" for special purposes while here on Earth. Certain old-timers and wise young-timers tell us that they didn't start living until they discovered the purposes for their lives and they have pursued them since -- with joyful exuberance. As long as they pushed that "yes" button to activate the chip, they found they could do amazing things, while also radiating great delight.

We call the spider's special purpose chip "instinct". Our chips are perhaps more complicated because we have more options. The spider's web-building actions are automatically programmed from its chip. One of our human options is to deny or ignore our chips. Another option is to use them in leisure or even in evil pursuits. As we explore our chips and search for meaning, may we do so in peace and empowerment for others.

~ ~ ~ ~ ~

READER SCORE SHEETS
To Score Places You Visit

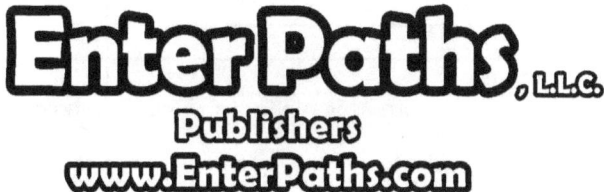

Copy this page and use it for your own scoring of places.
This one uses Clue sub-subjects as reviewed previously.

Your Example-1:

A = presence of item, 1 - 5 scale; **B** = weighted magnitude, 1-3 scale

1 - History Density

	A	x B	= Score
Strong sense of history in this place			
A connection with me I can't explain			
Feelings that I want to stay here a while			
Urgings to be quiet and not talk			
Feelings of something very special here			

Total score, this place, History Density =

2 - Intrinsic Symbols

	A	x B	= Score
Space - identification			
Light - present/contrasted			
Proportions - intrinsic			
Shapes - intrinsic			
Color - intrinsic			

Total score, this place, Intrinsic Symbols =

3 - Cultured Symbols

	A	x B	= Score
Stories anchored to this design			
Stories behind forms and shapes			
Stories behind layout			
Stories - accessories & furnishings			
Stories behind colors used			

Total score, this place, Cultured Symbols =

4 - Bio-Kinetics

	A	x B	= Score
Motion - foreground, background			
Motion in alignment			
Motion between angles, curves			
Motion as one moves around & through			
Motion in transparency			

Total score, this place, Bio-Kinetics =

5 - Intention

	A	x B	= Score
Of its being & purpose			
Expressed in shapes-form-placement			
Expressed color and honor			
Expressed in Bio-Kinetic dialogs			
Expressed in continuity			

Total score, this place, Intention =

READER SCORE SHEETS
To Score Places You Visit

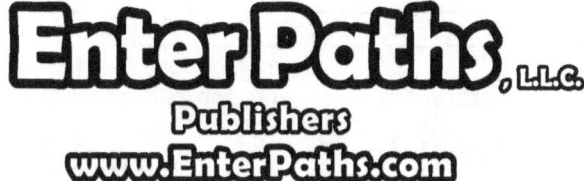

Copy this page and use it for your own scoring of places.
This one leaves the Clue sub-subject lines blank.

Your Example-2: _____

A = presence of item, 1 - 5 scale; **B** = weighted magnitude, 1-3 scale

1 - History Density

	A	x B	= Score

Total score, this place, History Density =

2 - Intrinsic Symbols

	A	x B	= Score

Total score, this place, Intrinsic Symbols =

3 - Cultured Symbols

	A	x B	= Score

Total score, this place, Cultured Symbols =

4 - Bio-Kinetics

	A	x B	= Score

Total score, this place, Bio-Kinetics =

5 - Intention

	A	x B	= Score

Total score, this place, Intention =

QUICK ORDER FORMS
For Free Information and Book Ordering

Quick Order Form

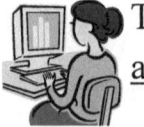 To order additional copies of <u>Thin Places</u>
<u>and Five Clues in Their Architecture</u>

Go to <u>://www.EnterPaths.com/</u> and follow links.

To request FREE information, send an <u>e-mail</u> with
your request including the information below to:
<u>@EnterPaths.com</u>

Please send Free information on:

☐ Other Books ☐ Speaking/Programs

☐ Our e-mailing list ☐ Downloads Available

Name: _____

Address: _____

City: _____ State: _____ Zip:_____

E-mail address: _____

QUICK ORDER FORMS
For Free Information and Book Ordering

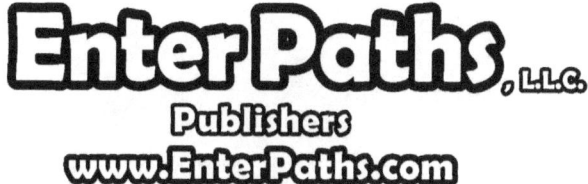

Quick Order Form

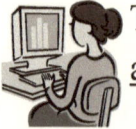 To order additional copies of <u>Thin Places</u> <u>and Five Clues within their Architecture</u>

Go to <u>://www.EnterPaths.com/</u> and follow links.

To request FREE information, send an <u>e-mail</u> with your request including the information below to: <u>@EnterPaths.com</u>

Please send Free information on:

☐ Other Books   ☐ Speaking/Programs

☐ Our e-mailing list   ☐ Downloads Available

Name: _____

Address: _____

City: _____ State: _____ Zip:_____

E-mail address: _____

QUICK ORDER FORMS
For Free Information and Book Ordering

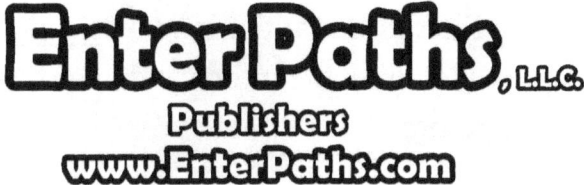

Quick Order Form

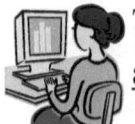 To order additional copies of <u>Thin Places and Five Clues within their Architecture</u>

Go to ://www.EnterPaths.com/ and follow links.

To request FREE information, send an <u>e-mail</u> with your request including the information below to: @EnterPaths.com

Please send Free information on:

☐ Other Books   ☐ Speaking/Programs

☐ Our e-mailing list   ☐ Downloads Available

Name: _____

Address: _____

City: _____ State: _____ Zip:_____

E-mail address: _____

www.ingramcontent.com/pod-product-compliance
Lightning Source LLC
Chambersburg PA
CBHW030751180526
45163CB00003B/985